Risk Analysis and Governance in EU Policy Making and Regulation

Bernardo Delogu

Risk Analysis and Governance in EU Policy Making and Regulation

An Introductory Guide

Springer

Bernardo Delogu
Independent Researcher
Rome, Italy

ISBN 978-3-319-30821-0 ISBN 978-3-319-30822-7 (eBook)
DOI 10.1007/978-3-319-30822-7

Library of Congress Control Number: 2016935850

Printed on acid-free paper

This Springer imprint is published by Springer Nature
The registered company is Springer International Publishing AG Switzerland

To my beloved wife Aurora who has encouraged me to write this book.

Contents

About the Author

Bernardo Delogu is a former Head of the Risk Assessment Unit at the Health and Consumers Directorate-General of the European Commission, currently retired after completing a 37 years career at the European Commission where he has gained extensive knowledge and experience of European policies in the areas of health, the environment, risk assessment, the internal market and research. As a European Commission high official, he has been in charge of developing, negotiating and implementing EU policies and regulation, notably in the former Directorate-General Health and Consumers as Advisor for Research and Science, Head of the Risk Assessment Unit, Head of the Products and Service Safety Unit and previously as Head of Sector and Policy Officer in the Environment, Industry and Energy departments of the Commission. Trained as a nuclear engineer, Bernardo started his career as a nuclear safety analyst in industry. He is now devoting part of his time to independent research on science and policy analysis.

Chapter 1
Introduction

The EU Risk Policy This book is an introductory guide to those concepts and methods of risk analysis that are most relevant for the development and application of the European Union risk policies and regulatory measures. It focuses on health, safety and environmental risks. It covers risk assessment, risk management and risk communication as well as the wider concept of risk governance. The general aspects and problems of the EU risk regulatory policy are also addressed. The concept of risk is seen in this book in a multidimensional perspective, since public perception, cultural aspects, values and expectations play an important part in the public discourse on risk, alongside the factual data and the scientific assessment.

The competences of the EU are defined by the Treaty on the European Union (TEU) and the Treaty on the Functioning of the European Union (TFEU). Some of these competences are "exclusive" meaning that only the Union may act, whereas others are shared with the Member States, in which case the Member States may exercise their competence to the extent that the Union has not exercised its own. In certain areas, the Union may instead only carry out actions to support, coordinate or supplement the action of the Member States.

Health and environmental issues play an important part, directly or indirectly in the development and application of several policies that fall under EU exclusive or shared competence or where the Union actively supports Member States' action. The Treaty states that environmental protection requirements must be integrated into the definition and implementation of the Union's policies and activities (Art. 11, TFEU) and that a high level of human health protection shall be ensured in the definition and implementation of all Union policies and activities (Art. 168, TFEU). Moreover, the Treaties prescribe that the European Commission in its proposals concerning health, safety, environmental protection and consumer protection, aimed at the establishment and the functioning of the internal market, will take as a base a high level of protection taking account in particular of any new development based on scientific facts (Art. 114, TFEU).

B. Delogu, *Risk Analysis and Governance in EU Policy Making and Regulation*,
DOI 10.1007/978-3-319-30822-7_1

EU policy and regulatory measures related to health and environmental protection are intended to address the relevant risks, preventing, reducing or controlling them. Therefore, the EU institutions play an important role as "risk manager" and risk decision-maker in many policy areas. Ensuring a high level of protection, as mandated by the Treaties, involves identifying, assessing and managing potential risks in an effective, efficient and proportionate way. Often, protection issues are characterized by uncertainty and EU decision makers are bound to act with full consideration of the precautionary principle. For example, EU institutions decide which pesticides may be used, in which way and concentration, account taken of their intrinsic toxicity, their fate in the environment, the routes of exposure etc., all aspects characterised by complexity and uncertainty. Which ingredients are acceptable in cosmetics, whether nanoparticles or nanostructures may be included in medical devices, which genetically modified organisms may be safely released into the environment, whether lighters must be equipped with child-resistant devices, which level of global warming is affordable without irreversible disruption of the environment, are just a few additional examples of the very diverse risk-based decisions that are taken on the EU-level and require consideration of complex scientific, technical, economic and societal aspects. EU decision makers are indeed faced with a challenging task when addressing health, safety and environmental risks due to the variety and complexity of the risk issues posed on the EU-level and the diversity of situations, interests, priorities and cultures across the Member States.

The EU Institutions (the European Parliament and the Council on proposal from the European Commission, or the Commission itself in cases where the EU legislation foresees delegated or implementing measures) act therefore as risk regulators in very diverse areas, like for example food safety, the safety of cosmetics, chemicals, toys, electrical products, medical devices, medicines, and on the full range of health, safety and environmental topics, for instance climate change, atmospheric pollution, control of communicable diseases, cancer prevention, etc. The EU decision-making process is itself complex and involves a plurality of actors: the European Commission, the Parliament and the Council, under the judicial review of the European Court of Justice and with the scientific and technical support of specialized agencies and bodies. Most of the risk issues addressed not only involve complex scientific and technical contents, but also have often far-reaching economic and societal implications. Moreover, EU regulation restricting the marketing and import of products or technologies has an impact far beyond the EU jurisdiction and must comply with international trade rules. The EU Member States have often different priorities and perceptions about risk issues. Dealing effectively with risks is not only very complex on legal, scientific and procedural grounds. It also implies addressing perceptions, societal preferences, values, and ensuring as far as possible an active and constructive involvement of a variety of stakeholders.

EU health and environmental policies and measures have been established over a long period of time, across many different sectors and do not follow a single, harmonised approach. The very concept of "risk" and the risk analysis paradigm do not play in the same way across the various policy areas. Nevertheless, the concept and methods of scientific risk assessment are central in all risk-related EU policy and regulatory processes, based on the assumption, sometimes naively interpreted as we will see, that protection measures should be "science-based".

Aims of This Book There are plenty of books, papers and academic articles on risk analysis and risk governance in general as well as on the more specific scientific and technical aspects of risk assessment and its methods. Some are quite general while others go into technical details in such a way and with such a language as to be of interest for the specialist only. Most of the existing books are written on the assumption that the separation of roles between risk assessors, risk managers and political decision-makers implies a corresponding differentiation in the need for access to the underlining knowledge. In particular, risk assessment methods are assumed to be of interest only for the specialised practitioners who have the tasks of developing and applying such methods and of informing policy decisions on specific risk issues by "translating" into lay language and explaining to policy makers the response of "science" to their questions.

The author of this book, in light of his experience on the EU decision making process, takes instead the view that decision makers, stakeholders and all the actors involved or interested in the EU risk policy and regulation would benefit from a deeper insight into the process of risk assessment, which is a pre-condition for a constructive, informed and fruitful risk dialogue and for successful and balanced regulatory decisions. The book tries in particular to show that risk assessment is a science-based decision-aid tool of vital importance for informing and supporting decisions, which however involves assumptions, uncertainties, expert judgements, conventional defaults, subjective probabilities, extrapolations and models that limit its validity, require qualifications and should be understood by all the relevant actors in the process and jointly addressed by managers, decision-makers and scientists. There is in fact a need for a transparent and shared "risk assessment policy" on important aspects like, for example, the criteria for determining tolerable, acceptable or negligible risks, the use of safety factors, the criteria for weighing scientific evidence etc. Such a risk assessment policy cannot be established solely on scientific considerations. Decision-makers and risk assessors should establish and maintain a systematic dialogue on the potential, limitations and meaning of risk assessment methods and results, as well as on risk assessment policy matters. Policy makers should therefore be more closely involved in framing risk assessment with a view to improve its utility for decision making and risk governance. The quality of the risk dialogue and the EU risk regulation can only benefit from a better shared understanding of risk assessment and its methods, going beyond the broad generalities, and of the risk analysis paradigm and the issues and problems of risk governance and risk regulation. The assumption of this book is that it is feasible for all actors, irrespectively of their background, to reach, with a reasonable investment of time, a level of competence on risk analysis that allows a critical view on the main aspects of the process and facilitates an informed dialogue between decision makers, managers, stakeholders and scientists.

The objective of this book is therefore to provide all those interested in the EU risk regulatory activities, notably the actors in the EU decision chain, with easy access to the information and knowledge that may help understanding the main aspects of risk analysis, governance and regulation, and may facilitate an informed and competent risk dialogue. The specific novelty of this book is its bringing together the main aspects of risk analysis, governance and regulation in the perspective of an integrated approach and with the needs of the non-specialist actor, stakeholder or interested researcher or observer in mind. An important aim of the book is to contribute to the productive exchange and smooth collaboration at the interface between science and decision making.

The Programme of This Book The book aims at providing a guide to risk analysis, governance and regulation with a very specific focus on the EU institutional, legal and policy context.

The starting point of our journey is a discussion on the concept of risk in Chap. 2. The concept of risk is understood and used often in an intuitive way. However, there is often confusion between risks and hazards and on the relation between risk and uncertainty. The concept of risk is in fact more subtle and complex than usually realised. Scientists define risk in a rigorous way, which is central to the structuring of risk assessment methods. But, risk is not just a technical concept. Perception of risk depends on diverse factors. Perceptions play an important part in shaping the risk debate. It is important to understand the multidimensional nature of risk as both a technical concept and a societal and contextual construct.

Chapter 3 introduces the concept of risk policy and risk analysis, which includes risk assessment, risk management and risk communication. The risk analysis paradigm was first developed in the 1980s in the US and since then has played an important role in structuring the regulatory policy also in the EU and internationally. Chapter 3 examines how and how far the risk analysis paradigm has been taken into account in the EU legislation and practice. It also discusses the need for integrating the risk analysis functions, in order to ensure a sustained interaction between risk assessment and risk management, and the importance of going beyond a purely "technocratic" view of risk management. The recent refinements to the original paradigm of risk analysis are also discussed.

Chapter 4 is a guide to risk assessment methods. It focuses on the assessment of health and environmental risks. As far as health protection is concerned, are considered chemical, microbiological and radiation risks, all aspects of great relevance in important EU policy areas. The safety assessment of food and feed from genetically modified plants is also briefly introduced. Chemical risks are also considered in relation to ecological risk assessment. A section of Chap. 4 is devoted to the safety assessment of nuclear power plants, an important example of engineering risk assessment. Chapter 4 also discusses several broader issues of risk assessment, some related to chemicals, like for example the assessment of mixtures, of nanomaterials, and of endocrine disrupting substances, others more general, notably the weight of evidence approach and the issue of association and causality in risk assessment.

Chapter 5 introduces risk management, defined as the identification and assessment of policy options in a multi-criteria perspective, account taken of the appraisal of the risk assessment findings. It discusses the principles applicable to risk management within the EU legal framework: subsidiarity, proportionality, precaution. Are also considered: the way in which the balance risk-risk is taken into account, the role of risk-benefit and risk-cost evaluations, as well as the decision criteria on the acceptability of risk. Finally, are briefly introduced the principles and procedures of the EU Better Regulation approach.

Chapter 6 is devoted to risk communication and risk dialogue. Risk communication, the third pillar of risk analysis, under the current best practice is a concept that goes far beyond communicating and “explaining” risks. It covers a structured and sustained dialogue along all the risk analysis process, across the decision chain and with the stakeholders and the public. The EU relevant practices are presented with some details on the most developed procedures of communication and consultation and the best practices for a comprehensive stakeholder dialogue.

Chapter 7 discusses both the issues and problems of risk regulatory policy and the need and ways for a participatory risk governance approach. Are introduced in this Chapter the rationale and the objectives for the EU risk regulatory policy and is discussed the way in which the Better Regulation approach established by the European Commission aims to structure the EU regulatory action across the entire life-cycle of regulatory instruments. Are also discussed the role of science in regulation, the need for an adaptive approach to risk regulation, the concept of evidence-based regulation and the relation between regulation, innovation and competitiveness. The regulatory process must be integrated into a wider governance perspective. Risk governance in the EU includes the “government” of risks, but goes beyond the formal institutional procedure of decision making and also includes a participatory approach involving the “civil society”. Such an approach is of the utmost importance in order to enhance the quality, effectiveness, efficiency and social acceptance of risk decisions. It is not without its own difficulties, which are also discussed in light of experience.

Chapter 8 discusses the needs and possible directions for the improvement of the quality and utility for decision makers of risk assessment. Risk assessment, in spite of the role it has gained in the EU risk decisions, is affected by a number of weaknesses and its role would become more relevant with a better co-ordination and collaboration between assessors and managers, a more policy-relevant expression of risks, a clearer, more systematic and transparent expression of uncertainties. The pre-condition of improvements is however, more clarity and a strict discipline on the respective roles of scientists and decision makers.

Finally, Chap. 9 synthetises all the main risk analysis issues with the relevant set of best practices, in light of the experience in the EU and the practices in other jurisdictions. The objective is to offer synthetic guidance on the key characteristics of well-managed risk analysis and governance of the regulatory process. This Chapter should help identifying possible weakness and limitations in specific risk-policy cases, on the level of both assessment and management. The proposed guidelines may also help identifying needs and possible directions for improvement in the current risk analysis structures, methods and practices.

Before starting our journey into the complex world of risk analysis, a final comment and a disclaimer are necessary. Most books and reports on risks, risk assessment and risk analysis have a very technical and sometimes academic profile and develop in detail one or more of the relevant issues. The characteristic of this work is instead to bring into a single, short book a variety of information and considerations on the wide range of issues that may help whoever needs or is interested in understanding how risk analysis comes into play in the EU policy and regulatory decisions on risk. Although the approach to the scientific aspects is rigorous, we aim in this book to use an accessible language, avoid unnecessary details and keep the book readable. We have limited the bibliographic references in the text, even if on each of the issues presented here the author has consulted much of the rich literature that is available. A short summary of the main issues covered is included at the end of each Chapter. The reader will notice that certain issues are mentioned several times in different Chapters and discussed from slightly different angles, and some statements are frequently repeated across the text. In that way, we try to facilitate the reading or consultation of each Chapter independently from the rest of the book, and to highlight some key messages of the book. Finally, the information and considerations presented in this book are very much based on the experience and knowledge that the author has gained in his years as official at the European Commission. However, the responsibility for the content of the book, notably any opinion expressed, obviously is exclusively his own.

Chapter 2
On the Concept of Risk and Its Application in Risk Analysis

Abstract The concept of risk is complex and multidimensional. It involves "objective" aspects, individual perceptions and societal values and preferences. Risk evokes possible damage or loss of utility as well as uncertainty about a future outcome. Although risks are often associated with potential benefits or opportunities, the negative connotation prevails in the common language. A technical definition of risk is at the core of regulatory risk assessment: risk is a function of the hazard in presence and the probability of the hazard to materialise and produce harm under the relevant specific conditions. An important distinction must be made between hazard and risk. Hazard refers to the intrinsic capability of an agent, event, state of a system, course of action, product, service, activity or technology etc. to cause an adverse impact on something considered valuable. Such adverse impact may or may not materialise, depending on the specific circumstances. Risk includes both hazard and the probability of the negative outcome. Confusion between the two concepts may lead to inefficient risk management measures. Risk perception plays an important part in risk governance. Risk is also a societal construct that depends on construed cultural and value-laden categories and paradigms. Perceptions and the position on risk of stakeholders and the public depend on contextual factors like for example degree of control, choice or imposition, benefits, familiarity etc. Risk can often be expressed only in qualitative terms and cannot be reduced to a single figure. It is by definition a probabilistic concept involving both an intrinsic degree of uncertainty and uncertainties affecting its determination.

2.1 Introduction

Since this book is an introduction to risk analysis, it is quite obvious to start it with a discussion about the concept of risk and its application to risk analysis. We will show that although the core of the concept of risk resides in the objective potential for a damage to something valued by humans (individuals, groups or the society as a whole), other dimensions and attributes intervene in determining how risk is conceived and construed by people and how it becomes a relevant societal and political fact.

There are many different types of risks, notably depending on their source (what is their cause, for example natural or manmade), their nature (what is at stake, for

B. Delogu, *Risk Analysis and Governance in EU Policy Making and Regulation*,
DOI 10.1007/978-3-319-30822-7_2

example health, environmental goods, financial investments, security) and their characteristics (irreversible, catastrophic, global, systemic etc.). However, it is possible and useful to identify and discuss some common, overarching aspects that intervene in a general definition of the concept of risk and in its characterisation across the different sectors and categories.

Many different types of risks are today addressed and regulated within the framework of EU-level policies. Although our introduction to the concept of risk in this Chapter is quite general, in the rest of the book we focus on environmental, health and safety risks since discussion of the assessment, management and governance issues of other categories of risks, for instance financial, security or strategic risks, also relevant on the EU-level, would require consideration of sector-specific contents and issues which are too heterogeneous to be dealt with in any significant depth all together.

Understanding and properly addressing risk require clarity about its component concepts and rigour in the use of the corresponding terminology. We will in particular discuss the important difference in risk analysis between risk and hazard, which is often overlooked and is a source of controversy, misinterpretation, confused debates and sometimes inefficient measures. Also, while uncertainty is intrinsic in risk, we will highlight that there are substantially different types of uncertainties that need to be addressed in risk analysis: the uncertainty *intrinsic* component of the probabilistic nature of risk on the one hand and the uncertainty *about* risk on the other hand, an important difference leading to different decision conditions: decision under risk and decision under uncertainty.

An overarching, technical definition of risk helps understanding the core of the EU approach to risk regulation. However, such general concept is in practice applied in different ways, depending on the sector. For example, while risk is defined and applied as a quantifiable concept in nuclear safety practice, a more qualitative approach often applies in other areas, like in particular product safety. We will see that the underlining concept of risk is hardly recognisable in certain regulatory instruments that require the regulated products to be "safe". Nevertheless, for policy and regulatory needs, criteria for grading risks or distinguishing between tolerable, acceptable and unacceptable ones are necessary.

We will see that risks cannot be effectively conceived and managed in a purely "technocratic" perspective. Public perceptions and attributes of risk, as seen in the perspective of those exposed, concur in defining and characterising risk-issues within open and accountable management and governance processes. Sometimes, the approach to risk is determined by political considerations, more than technical concepts.

We will highlight that "risk" is also a societal construct and is perceived, characterised and weighed by those exposed within a mental framework of individual or societal values and against its actual or perceived attributes. We refer in this Chapter to the respective importance, role and limitations of the technical, "objectivistic" concept of risk and of risk perception, a subject that will also be addressed at greater length further on in the book.

The concept of risk is often specular to that of opportunity or potential benefits. Without going as far as blurring the two aspects, it is important to highlight that the negative connotation of risk should not hide the link between risk, innovation and progress. The challenge for policy makers is in fact to prevent or mitigate the potential negative outcomes while safeguarding the intended utility and benefits of technologies, products or projects.

Finally, we briefly describe in this Chapter how the concept of risk is defined in the EU law and in some other jurisdictions.

2.2 The Different Meanings of Risk

2.2.1 The Multidimensional Nature of Risk

The Intuitive Concept and Common Definition of Risk It may seem that the concept of risk could be taken for granted, with no need for much discussion or specific definitions. In fact, it is a much more complex and multidimensional concept than it may appear at first glance. Thousands of pages have been written on how "risk" is conceived by the public, or should be defined for the objectives of governance. The concept of risk is indeed multi-dimensional and context-dependent. In fact, as we will see, there are several different conceptions and schools of thought on risk, highlighting its different dimensions.

In the common language the term risk is often used with the intuitive, qualitative meaning of something undesirable, which may or may not happen, but gives reasons for concern. The concept of risk is therefore associated with a possible adverse outcome, which is possible but not sure. "Risk" evokes possible "damage" or possible loss of "utility" as well as "uncertainty" about a future outcome. We talk of the risk of getting cancer from smoking or the risk of catching a flue from contagion in case of an epidemic; the risk to lose money investing in a hedge fund or to lose our job in an economic recession; or the risk to be late at work because of a possible traffic jam.

In the mentioned examples, the possible negative outcomes are of very different nature. The sources of the risk may be very diverse as may be diverse the targets exposed and the goods at stake. Nevertheless, the key conceptual elements characterising a "risky" situations are similar: the actual or even just perceived presence of a possible but uncertain negative outcome.

Dictionaries offer various definitions of risk. They highlight the chance or possibility of danger, loss, injury, or adverse consequence or the person or thing causing a risk, the possibility of something bad happening or what bad may happen. It appears from the most usual definitions that the word risk may refer, depending on the context, to the negative outcome, its probability or its cause.

Risk and Opportunity Some authors suggest that risk should in fact not be solely associated with a negative outcome but also with the concept of "opportunity" since human beings often accept or even look for risks in the hope of gain or other benefits. Typical examples are gambling games or poker, where the risk of losing money is accepted by players in view of (hypothetical) gains. In fact, mankind would have made little progress without its preparedness to accept reasoned risks in view of potential benefits. Just think of the geographic explorations that have open the way to new regions of the world. They would not have been possible without brave sailors accepting to risk their lives to verify hypotheses or uncertain information on the shape of the earth, the presence of undiscovered territories beyond the oceans or new routes to faraway regions, hoping to gain from any such discovery. While the overarching aim of risk analysis is to effectively and efficiently protect against risk as a negative fact of life and of human activity, one should not lose sight of the positive dimension of risk as a necessary factor of innovation and progress.

Nevertheless, although the perception of the link between risk and opportunity is undeniably at the basis of human behaviour in many circumstances, most frequently risk is perceived and understood as the distinctly negative side of the equation. Therefore, we mainly focus in this book on the negative connotation of risk, although consideration of benefits and opportunities will play an important part, later on, in our discussion on risk acceptance and risk management.

The Psychological and the Technical Perspectives of Risk We will see that much attention is given in the literature to the psychological or value and context-dependent aspects of risk, which should indeed be considered carefully in a comprehensive risk appraisal, and are very important for properly framing and addressing the risk-issues and for effectively managing the risk governance process. However, in the technical perspective of risk, which plays a central role in the assessment process, the core of the concept is a potential adverse event taking place in the physical world, consisting of an outcome causing a well-defined, real damage to something of value (health, ecosystems, property etc.), with a more than purely hypothetical possibility that the adverse outcome materialise.

Experience shows that such physical reality of risks materialises in a very tangible way when things go wrong. For instance, the Fukushima nuclear disaster shows that there was something very real and objective in the risk of an accident there following an earthquake, something dramatically more concrete than a subjective, psychological fact. It is clear, retrospectively, that the risk of an accident at the Fukushima site could not have been reduced to a "state of mind". There were very real conditions in the physical world (unfortunately not adequately identified and assessed) configuring a situation of risk: a site in a seismic region prone to tsunamis, inappropriate design and protection of emergency equipment, inadequately dimensioned barriers against flooding from the sea etc.

Risk Perception and Risk as a Social Construct In order to consider a risk real, the adverse outcome must have a reasonable degree of plausibility in light of the available knowledge and data. In the technical perspective, risk is conceived as an attribute of physical events with a probability different from zero to occur. However, it is important to consider that people who believe or fear to be exposed to a risk perceive, understand, weight and react to the feared potential damage in ways that depend on their position, interests, culture and values. Some people perceive a significant risk where scientists conclude to the absence or implausibility of a potential for damage or to a negligible probability of a damage to materialise.

It is therefore important to consider the psychological, subjective, contextual and societal dimensions of risk, notably its "social construct" dimension. "Social construct" may be defined as a conception, an opinion, a perception, a conceptual paradigm or a shared attitude that is construed through a collective, culturally or socially determined convergent, informal practice.

There are many examples of conceptual schemes that arise through collective thinking and assume a conventional reality going much further than their factual basis: typical are the stereotypes related to gender or racial characteristics

Various schools of researchers have explored the concept of risk from different angles (Renn 1992). We will discuss the relevance of the various dimensions of risks further on in the book. For now, let's mention that the "psychometric" approach focuses on the perception of risk (Slovic 1987), while the "contextualist" perspective highlights the importance of the attributes of risk, like for example whether it is voluntary or imposed, known or unusual, controllable or not etc. The "constructivist" approach stresses that the social context in which risk issues arise and management decisions are taken determine which dimensions of risk are most important. All such broader aspects and perspectives of risk are indeed important for a comprehensive risk appraisal and should be considered in implementing a participatory approach to risk governance.

Risk managers and policy makers need to understand and to take into account how stakeholders and the public perceive and frame specific risks within their particular perspective, values and interests, under the circumstances given in the specific case considered. It is also important to understand that there could not be in the real practice of risk analysis such thing as a purely "scientific" approach to risk since, as we will discuss later on, the "technical" risk assessment itself necessarily involves assumptions, choices, professional judgement, which are to some extent value-laden and inevitably bring in the process a certain degree of subjectivity.

2.2.2 The Technical Definition of Risk

The Importance of the "hard facts" of Risk Although the subjective and societal dimensions of risk play an important part in risk governance, the starting point of the well-established and internationally recognized risk analysis practices is the identification, assessment and, as far as possible, quantification of the "hard facts" of risk, those that can be objectively defined, verified and possibly quantified in appropriate metrics. In addition to appearing in the relevant definitions and requirements of the EU legislation, which we will present further on, the objectivist concept of risk is also internationally recognised. For instance, under the World Trade Organisation Agreement on Sanitary and Phytosanitary Measures,[1] measures taken by members must be based on risk assessment that involves (a) Identifying the adverse effects; and (b) If any such effect exists, evaluating the potential of occurrence (WTO 1994).

The technical concept of risk is therefore central in the traditional risk analysis paradigm. It defines risk as a function of the probability of an event and the severity and size of its adverse consequences. In that perspective, a risk is composed by the hazard involved, the likelihood of adverse effects to materialize as well as their gravity and size:

$$Risk = Probability\ of\ an\ Outcome \times Severity\ and\ Size\ of\ Adverse\ Effects$$

This definition is conceptually relevant across very different types of risks and activities. In some cases this definition allows the quantification of risk in terms of expected damage. For instance, if a certain population is exposed at a given dose-level to a carcinogenic substance over a long period, risk assessors may be in a position, under certain conditions, to calculate the number of expected additional cases of cancer over lifetime in the exposed population, indicating the degree of confidence in the estimate.

However, it is not always easy or even just possible to apply this definition in quantitative terms and with appropriate metrics. Moreover, comparing mechanically the levels of risk of different options may hide the underlying implications of risk. In fact, under this definition, a low probability event leading to a very severe damage or negative consequence, may be seen as involving a risk equivalent to that of a higher probability event causing a less severe negative outcome. This is a significant limitation of a quantitative approach based on the said definition.

As an example, let's imagine that a general is considering two possible ways to conduct a military operation involving 100 soldiers. Under option 1, there is a 20 % probability to lose 40 soldiers. The risk of option 1 is therefore $0.2 \times 40 = 8.0$, an expected loss of 8 lives. Option 2 involves an 80 % probability to lose 10 soldiers. The risk of option 2 is $0.8 \times 10 = 8.0$, again an expected loss of 8 lives. If the risk is evenly distributed, each of the soldiers involved in the operation has an 8 % probability of dying in both cases. In this example, clearly over-simplified for the sake of the demonstration, the two options have therefore the same level of risk, but clearly remain quite different.

Some authors suggest that the technical definition of risk should more properly be formulated as probability *and* consequences or a generic "function" of the two,

[1] The SPS agreement is a trade agreement that regulates conditions for international trade of agricultural, food, animal and plant products. It sets a discipline on the establishment and application of sanitary and phytosanitary measures aimed at food safety and animal and plant health. Under the agreement, governments are committed to take and apply sanitary and phytosanitary measures only to the extent necessary to protect human, animal or plant life or health and must not arbitrarily or unjustifiably discriminate between Members where equivalent measures are applied. The EU is bound to comply with the requirements of the SPS Agreement.

instead of probability *times* consequences. Moreover, we will see that for the policy and regulatory applications risks are expressed often in qualitative terms, with a terminology and grading that depend on the specific nature of the risk and the objectives of the legal framework applicable.

Whatever the definition adopted, it is clear that in most cases a risk cannot be reasonably reduced to a single figure. Comparing risks needs a careful consideration of all their aspects and dimensions: scenarios, probabilities, nature, implications and consequences of the possible outcome, etc.

2.3 Hazards and Risks

2.3.1 Definition and Categories of Hazards

The Distinction Between Hazard and Risk We have seen that we talk of risk when in presence of the possibility of an adverse outcome that may occur under certain circumstances. Both the intrinsic potential for damage and the circumstances under which the damage may materialise are needed for the risk to become real. In order to clarify under which conditions a risk may arise, we need to introduce the concept of "hazard". Hazard and risk are terms often used as synonyms. However, they are given clearly distinct meanings when used in risk analysis. The difference between the two concepts is of major relevance for risk policy and regulation. Across the various available context-specific definitions, "hazard", or any equivalent term used, is meant to represent the intrinsic capability of an agent, event, state of a system, course of action or behaviour, product, service, activity or technology to cause a negative impact on something that is considered valuable, like for example injury or harm to health, damage to ecosystems, damage to property or goods, financial loss etc.

For example, the UK Royal Commission on Environmental Pollution defined hazard as (UK Royal Commission 1997):

> A property or situation that in particular circumstances could lead to harm.

There are several sector-specific definitions of hazard and some of them are presented later on in this Chapter. Most of them are consistent with the general, classical definition of the UK Royal Commission.

Types of Hazards Examples of some categories of hazards related to health, which are addressed in EU policies and regulation, are listed in Table 2.1:

Hazards may also be related to a potential for adverse effects on the environment, on consumer or worker safety, on property or finance, on social goods, security, strategic interests, human welfare, wellbeing or psychological integrity and mental health. The distinction between hazard and risk is of general relevance, across many diverse sectors.

Table 2.1 Examples of hazards

Chemical hazards	Physical hazards	Biological hazards
Carcinogenicity	Ionization potential (radiation)	Pathogenic potential (microorganisms)
Mutagenicity	Thermal stressor	Viral pathogenicity
Toxicity	Electrocution potential	Fungal pathogenicity
Irritation potential	Vibration and Noise generation	Etc.
Sensitization potential		
Corrosiveness	Etc.	
Etc.		

2.3.2 *From Hazard to Risk*

As hazards represent only a potential for adverse effects, once a hazard is identified and characterised several additional steps are required in order to determine whether, how, to what extent, and under which conditions, their potential negative impacts would occur under any given circumstance, as shown in Table 2.2 for several different hazards:

Conditions for a Damage to Materialise Depending on the circumstances of each specific case considered, the adverse impacts of a "hazard" may or may not materialize, may materialize to a greater or lesser extent, may involve a broader or smaller population or set of targets, may materialize in the shorter or longer term or for a shorter or longer period of time.

The likelihood of an adverse impact to materialize in the presence of a given "hazard", as well as its severity and size, depend on the relevant circumstances. The circumstances of a hazard may include for example the way in which the hazardous agent is incorporated into a product, how the product is used, the composition of the population exposed, the vulnerability of the groups exposed, the level and duration of exposure etc. (Table 2.3).

Knowledge and consideration of such circumstances are key for the effective and efficient prevention, reduction or control of the potential adverse impacts.

Exposure as a Determinant of Risk For various categories of risks, like for example those related to chemicals or radiation, the exposure to the source of hazard plays a key role in determining the existence, gravity and size of a risk. As it will be explained in Chap. 4 on risk assessment, a toxic chemical incorporated in a product in such a way that its use could not cause any exposure of the user does not represent a risk in the utilization-stage of its lifetime in spite of its intrinsic hazard (it may however pose a health or environmental risk in other life-stages like in particular the end-of-life, waste-treatment and disposal stage).

Risk Versus Hazard Sometimes, the difference between hazards and risks is not well understood and properly taken into account in risk management. However, such a difference has substantive implications for determining appropriate, effective

Table 2.2 From hazards to impacts

Hazards *(The intrinsic potential for harm/damage)*		**Targets, Populations, Assets at Risk**		**Impacts**
Examples		Examples		Examples
• Toxicity • Carcinogenicity • Flammability • Ionization • Pathogenicity • Terrorism • Seismicity • IT breakdown • Supply breakdown	Exposure to Hazard	• Consumers • General Public • Specific Groups • Ecosystems • Natural Resources • Property • Infrastructure • Reputation • Confidence • Legal Security	Vulnerability, Level of Exposure	• Mortality • Morbidity • Loss of Quality Adjusted Life Years • Environmental Degradation • Resource depletion • Loss of biodiversity • Damage to Property • Financial Loss • Law suites • Reputational damage
Hazard Identification and Characterization	Exposure Assessment	Target Identification	Target/Exposure Characterization	Risk Characterization

Table 2.3 Hazard, risk and damage

Hazard *The potential to cause harm/damage* **Risk Factors** *Exposure, vulnerability etc.*	**Risk** *The combination of the probability of harm or damage occurring and its severity*	**Harm/Damage** *Loss of or damage to something that society values (health, property, environment etc.)*

and efficient measures aimed to protect against the potential negative impacts. Lack of understanding, or of proper consideration of such difference is most often a source of confusion in the risk dialogue and may lead to inappropriate or inefficient policies and measures.

The technical assessment of risk, based on the definition of risk introduced above, is aimed at providing information on both the nature and size of the *potential* negative impacts and on the circumstances, conditions and related probabilities for the adverse impacts to materialize, as well as their distribution and timescales. Information on the presence of a hazard only signals a potential for a negative impact, but does not provide any indication of its probability to materialize. This is the reason why the risk-based approach, as opposed to the hazard-based one, is generally considered to be the "golden rule" for policy making and regulation. Nevertheless, there is a lively debate on the respective merits and problems of the risk-based and hazard-based approaches. In the real world of risk policy, decision makers need both a good understanding of the difference between the two concepts and some reasoned pragmatism. On the one hand overcautious approaches based on hazards may unduly compromise innovation and its potential benefits; on the other hand, insisting on consideration of exposure to hazards and probabilities of adverse event occurrence when there is no reasonable way to determine, quantify and/or control them may result in inadequate measures followed by serious losses or damages. We will discuss in Chap. 5 on Risk Management the situations and conditions under which hazard-based policies and measures may be justified or even necessary, while warning against wasteful abuses.

2.4 The Technical Concept of Risk

2.4.1 The Engineering Approach to Risk

Risk as a Scientific, Probabilistic Concept Now that we have introduced the concept of hazard and the distinction between hazard and risk, we can come back to our discussion about the concept of risk. We have seen that the "technical" concept of risk combines the hazard in presence, as characterised under the given circumstances, with the probability of the potential adverse effects of the hazard to materialize, again under the circumstances of the case considered. With different formulations, all the definitions provided by various regulatory instruments therefore conceive "risk" as compounding the probability of an event and the severity of its consequences. Such technical concept of risk is derived from engineering, notably the engineering approach to the safety of nuclear facilities. The US Nuclear Regulatory Commission, published in 1975 a seminal document on probabilistic risk assessment, the "Reactor Safety Study-WASH 1400" (US NRC 1975). Quantitative, probabilistic risk assessment was in fact developed initially for assessing the safety of nuclear facilities and other engineering systems. It is based on an "objectivist" concept or risk (risk as a property of physical systems). In this perspective, "risk is a purely scientific concept admitting complete characterisation and analysis through data collection and quantitative methods" (Schroder-Frechette 1991), and risk assessment is a "fact finding process". As mentioned above, some authors add, or even oppose to this conception a subjectivist, constructivist or "contextualist" concept highlighting the importance of risk perception and/or the social context (more on this subject below).

The "engineering" approach starts with the identification of the events that can lead to adverse outcomes, for instance faults of technical devices or human errors, then it aims at estimating their probabilities and, finally, at identifying the cascades of possible events leading to an adverse outcome and calculating their compounded probabilities. The final overall aim is to assign probabilities to the possible resulting adverse effects. The process is therefore "bottom up", from the analysis of component faults and their propagation, to the characterisation of the final risk. In the engineering approach, the operational definition of "risk" is what is calculated in this way.

Different Expressions of Risk The technical definition of risk should be regarded as a conceptual scheme that may be adapted and applied in different ways, depending on the characteristics of the specific sector and the risk-problem considered. In areas other than engineering of technical systems, probabilities and gravity are in fact seldom expressly quantified and uncertainties are quantified based on expert judgment and sometimes expressed as subjective probability distributions. Risks are often established and expressed in qualitative terms only. In certain areas, the practice is to express risks by the use of technical quantities. For example, in the case of chemicals, health risks from the exposure to dangerous substances are characterised, as we will see more in detail further on, by quantifying "margins of safety", which however often do not provide decision makers with information on the probability of the relevant health damages. Sometimes and where possible, risks of exposure to chemicals or other stress factors are expressed in terms of expected increased incidence of certain pathologies. Risks may also be calculated and expressed in actuarial terms, as in the case of expected injuries or deaths in traffic accidents in a given year, based on reasoned extrapolation from accident statistics.

Sometimes, risks are graded (as tolerable, acceptable, negligible etc.) against pre-defined specific criteria, which for chemicals may be related to dose-thresholds or margins of safety. Grading and expression of risk belong to a risk-policy area borderline between assessment and management where sustained dialogue and collaboration between scientists and decision makers is particularly needed.

2.4.2 Subjectivity in the Application of the Technical Concept of Risk

The Limits of a Scientific Approach to Risk The technical concept of risk plays an important role in the EU risk analysis approach. However, it could not be isolated from the other important dimensions of risk. Moreover, it should be recognised that in its practical application, values and subjectivity inevitably play a role. A systematic effort is needed to reduce the degree of subjectivity and to ensure full transparency on all non-scientific aspects involved in the application of the technical concept of risk (determination of the relevant aspects of a risk issue, assumptions and defaults used, criteria to select data and weigh evidence etc.). Finally, conceiving risk as a purely scientific concept should not lead to overlooking valuable knowledge from unconventional sources. We will come back in more detail to the latter issue when discussing in Chap. 8 how best to organise risk advice.

2.5 Risk in Practice

2.5.1 Relativity of Risk

The apparently straightforward technical concept of risk needs some important supplementary considerations, additions and qualifications in view of its use in practical applications.

Risk Distribution First, risks are rarely uniformly distributed across the exposed population. Risk cannot in most cases be reduced to a single figure or statement. For instance, in view of risk management decisions, is in general not helpful to provide managers with just an estimate of the average level of risk. Such average may be low, although a part, perhaps a small but not negligible part of the population is exposed to a significantly higher-than-average level of risk. The degree of exposure as well as the vulnerability of exposed groups and the resulting severity of the consequences of exposure can vary greatly across the population (or other targets) at risk.

Relativity of Risk Gravity Second, the value of the menaced "good" at stake, and therefore the gravity of the adverse impact, may not be the same for all the receivers of a risk. For example, owners and users of an office block in a certain part of a town may grade risks related to poor air quality and high level of noise due to unrestricted car access on a much lower level than permanent residents in the same zone: air quality and quiet streets are comparatively less important for commuters or office-block owners, who may value more unrestricted access by car to the office, than for permanent residents who live or have invested in residential buildings in the zone.

Risks and Benefits Third, in risk analysis consideration of risks cannot reasonably be separated from possible benefits and their distribution. This is particularly important when trying to grade risks comparatively. The health risks associated to a substance used in a life-saving medical device can hardly be compared, without considering its risk-benefit balance, with those of the same substance when used in toys or cosmetics.

2.5.2 Risks, Opportunities and Benefits

The Different Metrics and Distribution of Risks and Opportunities As already mentioned, some authors and organizations consider that risks and opportunities should always be regarded as two sides of a same coin or even just two aspects of the same concept. Although, as we have just noted, risk management may require careful consideration of a risk-risk and/or risk-benefit balance, blurring the distinction between the very concepts of risk and opportunity/benefits is not justified and does not help the assessment in practice since the metrics and distribution of risks and benefits are in general different and the balance between positive and negative aspects cannot reasonably be comprehensively expressed with the same units as a single value (in spite of the efforts and progress made by economists to express most of the health and environmental impacts and benefits in monetary terms).

Assessing Risks and Benefits and Deciding on the Appropriate Balance Decisions on the appropriate balance between risks and benefits, as separate aspects of an event, product or technology, may not be left to risk assessors since they involve consideration of societal values. Such decisions therefore remain in the realm of risk management and political decision. Nevertheless, in order to properly inform decision making, it is indeed important to organise the assessment process in a way ensuring the parallel, structured and systematic determination of both risks and costs and benefits (and of their respective distributions). This objective has been addressed for example within the framework of the EU legislation on chemicals (European Union 2006), by establishing two committees working in parallel, one charged with risk assessment and the second with the assessment of socio-economic aspects, including costs and benefits. The information from both risk and socio-economic assessments is fed into the final political-decision stage of the process. We will come back in more detail to the issue of risk-benefit balance in Chap. 5 on risk management.

2.6 Risk as a Social Construct

2.6.1 Risk Perception

Risk Perceptions as Facts From its beginnings, EU risk policy has gradually evolved towards a more participatory approach. The experience with risk-crises has shown that a purely technical approach to risk does not suffice to effectively respond to societal concerns and expectations. A wide and varied school of thought, critical to various degrees and in different ways of a simplistic application of the technical concept of risk, highlights the subjective and the societal or contextual dimensions of risk. In some extreme interpretations, risks exist only in the mind of people, and is shaped by perceptions, expectations and values. A more balanced view does not deny the reality and relevance of the objective aspects of risks, but focuses on the need to explore and take into account in a comprehensive appraisal supporting the risk management and governance processes the perceptions, experiences and values of people. In the policy-making arena, perceptions are facts.

Perceptions Cannot Replace Hard Evidence and Scientific Assessment The societal dimensions of risks are among the aspects at the core of the risk governance paradigm that will be introduced in Chap. 7. For now, we need to note that an extreme overstatement of the "psychological" dimension of risks would not be of great relevance in the regulatory practice given that under the EU legal order, as well as under international law and trade agreements, binding measures might obviously not be motivated on ground of purely subjective perceptions. Nevertheless, understanding risk perceptions as well as the factors that determine public positioning on risks may help in identifying aspects that need to be included in the assessment in order to respond to concerns of the public and contribute to a participatory approach within a rational framing of all the relevant issues.

While a conception of risk as a purely mental construct expressing emotions does not help much for an effective and efficient approach to risk, the social "constructivist" approach is closer to the practical needs of an effective governance practice. Societal concerns, values and priorities are indeed important in shaping and framing the concept of risk. In particular, although the physical aspects of risk do not depend on culture and values, defining what is at stake and deciding which aspects must be addressed in dealing with a risk-issue require the involvement of the civil society and the consideration of the underlying structuring values, interests and objectives. For instance, in the sensitive cases of novel food, food from cloned animals or GMO food, aspects like nutritional quality, national and local traditions, the integrity and shape of the countryside, animal welfare and even the cultural and socializing role of traditional food, all concur in shaping the attitude and expectations of the European public in relation to the relevant EU policy and regulatory decisions.

2.6.2 Risk Attributes Determining Attitudes Toward Risks

Although risk could not be equated with perceptions, and perceptions may not be taken as a basis for regulation, as we have just said, perceptions are nonetheless "facts" and carry much weight in the democratic policy making process.

The Psychometric Approach In the so-called "psychometric" approach, perceptions are considered as just emotional reactions that need exploring in order to determine ways and messages aimed to "explain" in lay language the scientific facts and educate the public. "Perceptions" are in fact sometimes dismissed by tenants of "sound science" as purely emotional and irrational. Nevertheless, public rejection of certain products or technologies are often wrongly attributed to a purely emotional risk "perception" while it is in fact based on an intuitive but far from irrational positioning vis a vis the possibility of a risk. For example, the position of a risk receiver in accepting or rejecting an even hypothetical risk may be influenced by the degree of control allowed in case the risk materialises, the possibility of choice (risk chosen versus risk imposed), the benefit for the receiver (what is in it for me), the familiarity (fear of the unknown or uncertain that may reserve surprises), degree of trust toward the risk promoters, experts, authorities (why should I trust them?) etc. All these attributes of risk are quite objective and it is not necessarily irrational for the public to factor them in the intuitive appraisal of risk, notably when in doubt on the materiality and characteristics of the risk.

Risk Aversion and Risk Acceptance Risk aversion and risk acceptance depend on both subjective attitudes and objective considerations related to potential benefits. Personality and circumstances intervene in determining the position in relation to a risk, including possible expected rewards from accepting the risk. Rewards may be concrete and material, as financial gain, or immaterial and psychological as in the case of people practicing extreme sports.

What can be called a "contextualist" concept of risk highlights some very relevant attributes of risk that help broadening the perspective of the risk issues. Some of the most relevant risk attributes, most often cited in the literature, determining the perception of risk (or better the attitude toward risk) are shown in Table 2.4.

2.7 Characterising Risks in View of Governance

Depending on their nature and on the available knowledge, risks present different characteristics that have a significant impact on the way they are seen by stakeholders and the public and on how they should be approached in the management and governance process. A useful characterisation in that respect has been proposed by the International Risk Governance Council (IRGC 2005). Although many risks are simple in their causal structure, well understood and non-controversial, making risk management and governance straightforward, difficult cases are characterised by complexity, uncertainty or ambiguity.

Complexity Complexity may be due to the interaction of many factors in the causal relationships leading to the adverse outcomes and determining their probabilities. Assessment of such complex risks requires the application of sophisticated modelling techniques, not easily understandable by decision makers and the public.

Table 2.4 Risk perception attributes

Risk perception enhancing attributes:	Risk perception mitigating attributes:
Imposed	Voluntary
Not controllable	Controllable
Without proportionate benefits for those exposed	Associated with proportionate benefits
Affected by uncertainties	Clearly defined
Unknown, new	Well known, familiar
Irreversible	Reversible
Assessed and managed by untrusted entities (companies, experts institutions)	Assessed and managed by trusted entities
Unfairly distributed	Distributed with equity
Potentially catastrophic	Limited in extension and consequences
Dreadful	Psychologically bearable

In principle, in such cases it is possible to determine the risk in a quasi-deterministic manner, but often in the absence of complete knowledge and specific data, models include simplifications and assumptions that limit their validity.

Uncertainty Different types of *uncertainties* intervene for a variety of reasons, weakening the level of confidence in the causal relationships, models, data, parameters used. That can lead to action based on precautionary considerations, sometimes addressing hazards rather than risks.

Ambiguity Finally *ambiguity* characterises risk situations where divergent values between the various interested parties play a major role in determining the divide of attitudes and appraisal perspectives, leading to the need to integrate such different views in a common, comprehensive framework.

We will come back to this categorisation further on in the book. For now, let's warn that although this categorisation provides a useful interpretative scheme, one should take into account that the nature of controversial risks is rarely so clear-cut. In many controversial cases all these characteristics of risk have, to a different extent, been mixed up.

2.8 Different Categories of Risks and Risk Grading

2.8.1 Different Categories of Risk

There are many ways of categorizing risks. Different categories may require the application of different assessment methods; in certain cases, the differences are mostly relevant in relation to policy making, perceptions and governance.

Types of Risk We should first note that there are as many types of risk as there are categories of material or immaterial things of value that can be lost or damaged under certain circumstances. Any list would of course be incomplete and we have already identified the types of risk on which this book focuses: health, safety and the environment.

An obvious difference exists between "***natural***" and "***manmade*** risks because of their origin and nature, the degree of human responsibility in their appearance and the feasibility of prevention, reduction and control measures.

Natural Risks Natural risks could not always be considered as just unpreventable or unmanageable "acts of God". In many cases, there is scope for preventative measures also against natural risks (like for example safety standards for buildings in seismic regions, appropriate siting decisions for nuclear plants in order to prevent or mitigate the consequences of natural events like earthquakes or tsunamis, alarm system and emergency plans, building of adequate sea-barriers and flood-prevention works etc.). Natural risks provide in fact a good example of the importance to dis-

Fig. 2.1 Risk and its components

tinguish between hazard and risk. If natural hazards are most often beyond human control, the same is not true for risks that depend on additional factors like location, vulnerability and exposure, aspects on which preventative measures are often feasible, as shown in Fig. 2.1.

Global Risks Certain risks are categorized as "global". In addition to the size of their potential impact, spanning most if not all of the globe, they are usually characterized by the indivisibility of such impact. When risks are global, the consequences of actions by any relevant actor anywhere are felt by all those exposed everywhere. Typical example of global risk is the risk of climate change associated with man-made emissions of greenhouse gases.

Systemic Risks A further category is that of "systemic" risks. Those are risks affecting systems on which the functioning of our society depends, such as for example energy, transport, communication infrastructures. Systemic risks by definition have cross-sector implications and cause impacts that may involve physical, economic, health, social consequences.

2.8.2 Risk Grading, Quantification and Expression

Under certain conditions, the technical definition of risk provides the basis for a quantitative approach to risk assessment (Kaplan and Garrick 1981). The assumption is that both the probability of an event and the gravity of its adverse impact could, at least in principle, be expressed quantitatively. Quantitative expressions of risks are in fact used in various sectors, with grading and expressions that depend on the nature of the risk considered. We have seen the example of the risk of major accidents in nuclear engineering, which are usually quantified in terms of both probability and public health consequences, using a probabilistic approach and the modelling of the public health consequences of a major accident.

Qualitative Risk Grading In practice, depending on the nature of the risk considered and/or the difficulty to assess it quantitatively, only a qualitative grading is presented in many cases to risk managers. Qualitative terms like "serious", "negligible", "high", "moderate", "low" risk are often used. Probabilities are also characterized qualitatively, with terms like "very likely", "likely", "unlikely" etc. In the absence of supporting definitions, such terminology may however be interpreted in very different ways and introduces a considerable degree of subjectivity and a high risk of inconsistency in the understanding and appraisal of the risk by managers and policy makers. Moreover, a qualitative characterization of risks involves subjective value judgments. In the absence of criteria and definitions endorsed by, or at least

Table 2.5 Risk grading

		SEVERITY OF HAZARD				
Probability of Occurrence	Grading	Negligible	Low	Moderate	High	Very High
	Very Unlikely	LOW	LOW	LOW	LOW	MEDIUM
	Unlikely	LOW	LOW	LOW	MEDIUM	MEDIUM
	As likely as not	LOW	LOW	MEDIUM	MEDIUM	HIGH
	Likely	LOW	MEDIUM	MEDIUM	HIGH	HIGH
	Very Likely	LOW	MEDIUM	HIGH	HIGH	HIGH

expressly shared with policy makers, the use of qualitative terms by the assessors may also introduce some confusion on the respective responsibilities of risk assessors and managers (who decides whether a risk is to be graded as serious, high, low, negligible etc.?).

An example of qualitative, generic grading of risk is presented in Table 2.5.

Tolerable/Acceptable Risks In health risk assessment of chemicals, exposure is often characterised as "tolerable" or "acceptable", (daily or weekly intake of a substance: for example tolerable daily intake TDI or acceptable daily intake ADI). Such terms indirectly refer to the underlying toxicological risk and are used for grading the risk corresponding to a given level of exposure. Such grading is based on conventional criteria used in risk assessment, as it will be shown when discussing later in the book the assessment of health risks. Again, the use of these qualifiers of risk poses the problem of responsibility for the definition of the risk criteria applied.

Expression of Risks

The ways in which risks are quantitatively or qualitatively expressed are often determined by the legal framework applicable to the product, technology or activity considered. Risk assessment reports provided by EU bodies for supporting EU decision-making often qualify the use of a substance, for example in cosmetics or food, as "safe" or "not safe" or as giving or not a "reason for concern" or posing or not posing a risk to health. Conclusions formulated in this rough way reflect the questions that are posed by managers to risk assessors, which in turn are determined by the expression of the safety requirements in the applicable legislation.

SCCS Opinion on Acid Orange 7

A good example, taken from a real case, is provided by one of the many opinions that a Scientific Committee of the European Commission, the Scientific Committee on Consumer Safety (SCCS), adopted on request from Commission services, in order to support decisions on the implementation of the EU legislation on cosmetic products (SCCS 2014). Most of the opinions delivered by the SCCS are quite technical and relate to specific ingredients used or proposed for use in cosmetics. In the case of the substance called "Acid Orange 7" the following questions were posed by the Commission:

(continued)

1. In light of the new data provided, does the SCCS consider Acid Orange 7 (C015) safe? At on-head concentrations up to 0.5 % under non-oxidative conditions?
2. Does the SCCS consider Acid Orange 7 (C015) safe at on-head concentrations up to 0.8 % under oxidative conditions?
3. Does the SCCS have any further scientific concerns with regard to the use of Acid
4. Orange 7 (C015) in cosmetic products particularly as it is used as colorant agent with the name of CI 15510?

It should not be considered surprising that Commission's risk managers asked the scientific committee to pronounce on whether the substance in question were "safe". The legislation applicable to cosmetics states that only safe products may be placed on the market. Commission services were therefore aiming at a clear and simple answer, providing a solid basis for ensuring the legal security of a possible measure. The SCCS replied with an opinion adopted on 18 June 2014, concluding that the use of Acid Orange 7 as a hair dye ingredient up to a final on-head concentration of 0.5 % under non-oxidative conditions did not pose a risk to the health of the consumer. The SCCS did not pronounce on the safety of Acid Orange 7 use in oxidative hair dye formulations noting that its stability under oxidative conditions was not known. Moreover, the SCCS stated that the safety of Acid Orange 7 as a cosmetic colorant could not be assessed because no information on the use concentrations of Acid Orange 7 as a colorant in other cosmetic products was available and the aggregate exposure was not known.

Acceptable Risk The terminology used in the questions and answers of the example above implies an oversimplification of the concept of risk. It is in fact rarely possible to reach a "scientific" conclusion of zero risk in real cases with the methods and data that are available. In the absence of pre-determined criteria, when scientists are asked by risk managers to pronounce with a "yes or no" statement on the safety of a product or technology, they are pushed in practice to decide how much safe is safe. In his book "Of Acceptable Risk" published in 1976, William W. Lowrance observed that safety should be defined as a judgement on the acceptability of risk concluding that something is safe if its risks are judged to be acceptable (Lowrance 1976). Obviously, the immediate questions are: judged by whom, and acceptable to whom? This book aims in particular to show that science alone cannot and is not entitled to answer such questions. Those are questions that can only be properly addressed through a structured collaboration between assessors and managers, balancing science, public participation and political decision, in a process where the political institutions maintain the responsibility for the last word, but participation of stakeholders and the interested public is ensured.

The subject of quantification, grading and expression of risks is one of those on which a critical and informed attention of policy makers and a close collaboration between assessors and managers are most important. It should be considered an area for further improvement of current practices.

Risk and Safety An important observation may be made at this point, in light of the considerations and example exposed above. This entire Chapter insists on "risk" as the core of risk analysis and risk assessment, and on uncertainty and probability as essential components of the concept of risk. Now, the observer of the EU policies might be surprised when reading the "risk assessment" reports of the EU bodies that many of them… do not refer to the concept of risk as defined here! In fact, depending on the relevant legal framework, the objectives of the consultation and the nature of the issue, often scientists are just asked to provide advice on the presence/absence or extent of adverse effects possibly caused by a given level of exposure to a stress factor, like in particular a dangerous chemical. In many cases, when chemicals are considered, the answer can be given extrapolating from the available toxicological data with relatively simple calculations.

For instance, in the case of a chemical whose toxic effects are considered, based on animal models, to show up only above a threshold-dose, when the exposure from the application considered is below the threshold, scientists will talk of "safe" use. Alternatively, they will advise on the "safe" concentration or exposure that may be translated into a legal limit value. However, also in such "simple" cases, uncertainties and probabilities are often hiding behind the toxicological data used and the assumptions on exposure which are made.

2.9 Risk, Probability and Uncertainty

2.9.1 Uncertainty in Risk and Uncertainty on Risk

We have stated several times in this Chapter that technical risk assessment is aimed at providing risk managers with information on both the severity and the likelihood of adverse impacts caused by a risk factor under certain circumstances. The probability of an adverse impact may range, depending on the hazard and the circumstances of the case, from 0 (impossible event; no such impact) to 1 (event certain; the negative impact always materializes in the presence of the hazard and under the conditions considered). Clearly, in the latter case, when the probability is 1 or very close to 1, the concept of risk is of no practical relevance because one is in presence of a virtually certain adverse outcome.

Uncertainty Inherent in the Concept of Risk A simple and extreme example that helps understanding the concept of risk and the way in which uncertainty is inherent in it, is provided by the so called "Russian roulette". Russian roulette is known as a stupid and irresponsible game of chance (its Russian origin is only alleged!). In this game, each participant in turn places a single round in a revolver, and after spinning the cylinder randomly, pulls the trigger with the muzzle pointed against his head. The hazard in presence here is death of the player. Normally, only one chamber is loaded, therefore, for a revolver that holds six rounds, the chance of death is one in six, on the (theoretical) assumption that the position of the loaded chamber is perfectly random in all trials.

The technical concept of risk is easy to apply in this example as both the hazard (death) and the probability (1/6) of the adverse outcome are perfectly determined and equally distributed across all participants (the exposed group). It is also easy to understand from this example how uncertainty is intrinsic in the concept of risk. The player is each trial in a state of uncertainty about the outcome of pulling the trigger, but he is in a position to deterministically quantify such uncertainty based on "a-priori" probability.

It has been observed that when there is a risk, there must be something that is unknown or has an unknown outcome and therefore it can be said that characterising risks involves quantifying or qualifying our lack of knowledge, in a sense building knowledge about the lack of knowledge.

In our example above, the player does not know whether the chamber of the revolver is loaded of not, but can at least quantify the chance (1/6), in other terms he can characterise quantitatively his lack of knowledge.

Uncertainty on Risk The uncertainty which is inherent in the concept of risk should not be confused with the uncertainties of different nature and level, which play an important part in most cases, which are often related to one or more of the various components of the risk considered. For instance, in our example we have made the assumption that the position of the loaded chamber be perfectly random, which requires in particular a cylinder perfectly equilibrated. However, there may be slight defects in the cylinder, tolerated in the manufacturing process. Such defects may affect the randomness of the position of the loaded chamber. In the absence of precise information about the influence of the defects on the randomness of each trial, they may result in a further level of uncertainty for the player about the reliability of the theoretical 1/6 death probability estimate itself. This example shows that it is therefore important to be clear about the difference between the quantifiable uncertainty intrinsic to risk and the uncertainties on the risk itself.

In most cases, risks are affected by uncertainties of various nature and often the probabilities cannot be quantified based on specific and reliable data or statistical evidence. Transparency about the nature and size of uncertainties is of great importance for risk management. Unfortunately, this is an aspect which is not yet consistently addressed and treated in terms of methodology, terminology and communication across different policy sectors by the various risk assessment bodies.

2.9.2 *Uncertainty and Other Limiting Factors in the Application of the Concept of Risk*

Uncertainty in risk assessment and management is discussed in more detail further on in this book. Nevertheless, we need to make some preliminary considerations here on the way in which uncertainty intervenes in the application of the concept of risk to risk assessment and management.

Sources of Uncertainty in Risk Analysis Uncertainty is everywhere in risk assessment: the models used may be affected by uncertainty, as well as the values attributed to the parameters introduced into the models and the input data used. And uncertainty may derive from the randomness of the underlying phenomena, from a lack of data or imperfect or incomplete scientific knowledge, or from the intrinsic unpredictability of future outcomes, as in the case of future developments determined by human decisions and behaviour.

In addition to proper uncertainty, risk assessors have also to deal with variability, which may or may not be treated in a deterministic way, depending on its causes and the data available. Uncertainty, variability and incomplete knowledge often limit the applicability of the technical definition of risk. In some cases it is simply not possible to provide decision makers with a "full" risk assessment.

Decisions Under Risk and Decisions Under Uncertainty In the presence of significant uncertainties, decision makers take decisions under uncertainty, as opposed to decisions under risk based on known probabilities. Risk assessment uses specific techniques to identify, deal systematically with, express and communicate such uncertainties, but in many cases, depending on the nature of the issue, the state of knowledge and the data available, assessors cannot resolve all of them. And of course the "unknown unknowns" remain by definition outside the radar screen of risk assessment. In the last instance, it is for the risk manager and the decision maker to resolve the remaining uncertainties and draw conclusions from the state of scientific knowledge presented by the assessors.

Together with uncertainty, several other limitations affect the application of the technical concept of risk in the assessment process. Two particularly important considerations should be made in that respect.

Divergent Expert Opinions on Risk First, as we have already noticed, the process of identifying and assessing risks involves conventional and, sometimes, subjective assumptions, expert judgments, simplified models, extrapolations of data and use of default values in the absence of case-specific information. Such limitations may lead to divergent conclusions by different groups of experts or individual assessors. Divergences may arise from the use of different data sets, different models, different weight of evidence criteria, different interpretation of data etc. In essence, as we will see in more detail later on, technical risk assessment is a science-based process, in the sense that it makes use of scientific knowledge, instruments, criteria and methods, but is not "pure" science.

The Influence of Divergent Interests, Values and Perceptions in the Application of the Concept of Risk Second, the identification and assessment of risks most often do not take place in a quiet, dispassionate environment. On the contrary, it happens in the middle of fierce arguments reflecting different objectives, values, economic or trade interests, political stances. The discourse on risk of the different stakeholders involved, their support or criticism for a scientific definition of risk, their acceptance or rejection, on various grounds, of the risk assessment process, the over-statement, endorsement or denigration of a risk-based approach, are often substantially determined by the respective objectives, agendas, interests, ideologies and values.

The limitations of the technical concept of risk and the theories on the subjective psychological and societal dimensions of risk may be sometimes instrumentally exploited to dismiss the hard facts on risk that are established by technical risk assessment. In such cases, the public debate on risks loses any scientific framing. The consideration of facts about the risk considered may get mixed up with considerations on values and societal preferences and views on decision criteria and processes. In order to prevent such counterproductive confusion, while allowing for the proper consideration of all the dimension of risks, it is important to establish a sustained risk dialogue and a robust and structured governance framework, as discussed in Chap. 7.

The various conceptions of risk and the related assessment and appraisal methods and techniques are most useful for supporting policy and regulatory decisions in an open, participatory and democratic process as long as all those involved in the risk governance process share a common view on their meaning and potential, but also on their limitations.

The objective of this book is to provide the basic information and the tools that may help understanding and managing efficiently the complexities of risk analysis and governance under the EU conditions.

2.10 Risk in the EU Law and in International Instruments

2.10.1 Risk in the EU Law

Risk in the EU Treaties The EU Treaties do not provide a definition of risk. In fact, the word "risk" is mentioned in the Treaties only once, without a definition, in Art. 196 of the TFEU in relation to civil protection. Nonetheless, risk and risk assessment are at the core of decision making and regulatory practice in many EU policy fields, notably in the areas of health, food safety, consumer protection, the environment and climate, as well as the harmonization of technical rules for the objectives of the internal market. However, the concept of risk is not univocally defined and codified across the various EU policy areas.

Sector-specific definitions of risk are included in certain instruments of secondary EU legislation, with somewhat different wording and with a scope limited to the domain covered by the specific act.

Definitions of Risk in the EU Law The available definitions of risks reflect a common technical concept that includes, with different wording, the core elements of hazard and probability. Examples are:

- Regulation (EC) N° 178/2002 (the EU Food Law), Art. 3.9 and 3.14, defines risk and hazard as follows:
  - Risk means a function of the probability of an adverse health effect and the severity of that effect, consequential to a hazard.
  - Hazard means a biological, chemical or physical agent in, or condition of, food or feed with the potential to cause an adverse health effect.
- Directive 2009/48/EC on the safety of toys defines risk, hazard and harm:
  - Risk means the probable rate of occurrence of a hazard causing harm and the degree of severity of the harm.
  - Hazard means a potential source of harm.
  - Harm means physical injury or any other damage to health including long-term health effects

Definition of Risk by the European Commission The European Commission has provided a horizontal definition of risk in a "Toolbox" document accompanying the recently (May 2015) adopted new Impact Assessment (IA) Guidelines (European Commission 2015). In this document the Commission highlights the distinction between hazard and risk:

> A function of the inherent properties of the agent/event in question-whereas risk is: A function of both the hazard and the potential likelihood and extent of being exposed to the hazard.

The Commission states that

> In other words, while hazard represents an abstract danger, risk expresses the combination of the level of hazard and the likelihood of its occurrence. Risk = Hazard (expressed in terms of its negative impact) x Likelihood of its Occurrence.

The Commission notes that

> while the two variables are not independent from each other and while the impacts of the hazard depend on preparedness or preventive behavior (as in the case of natural hazards), the risk must be expressed as a functional relationship rather than a simple multiplication of the two variables.

The interest of this definition resides in its wide scope, its level of generality and the role that it may play through the application of the EU Regulatory Impact Assessment requirements across a wide range of proposals and decisions. Nevertheless, some refinements and a more accurate choice of terms[2] would be useful in view of clearer guidance and a coherent application.

2.10.2 *Definitions of Risk on the International Level*

Some Definitions of Risk on the International Level On the international level, various relevant definitions have been established

- ISO defines risk in its Guide 73-2009 "Risk Management-Vocabulary" as:

> The effect of uncertainty on objectives

where objectives can have different aspects (such as financial, health and safety and environmental goals and the risk is often characterized by reference to potential events and consequences or a combination of these (ISO 2009). The Guide states that risk is often expressed in terms of a combination of the consequences of an event and the associated likelihood of occurrence.

- Codex defines both hazard and risk as (FAO 1997):

> The generally accepted definition for the term "HAZARD" is a biological, chemical, or physical agent in, or condition of, food with the potential to cause an adverse health effect. It can be an agent of biological origin (such as pathogenic bacteria, viruses, parasites, or other organisms) or the product of these organisms with properties which can harm human health. It can also be a chemical substance in or on the food from external sources such as industrial pollution, agricultural inputs of fertilizers or pesticides, or substances added to food intentionally or by accident in excess

(continued)

[2]Although the overall intended meaning is clear to those knowing the subject, it may be confusing in a guidance document to refer to hazard as "an abstract danger". In addition to the fact that hazard and danger may be understood as synonyms, the intrinsic properties that constitute a hazard are far from being "abstract" (think of carcinogenicity, flammability, proximity to an active volcano…). There is a risk to induce the idea that hypothetical properties may be considered a hazard, when hazard identification and characterization should instead be based on a rigorous approach. Also the term "potential likelihood" would deserve some refinement or clarification. Generally, it is intended to refer to the actual likelihood of the potential adverse effects (the hazard) to materialize under the conditions and circumstances given.

of safe levels. It may also be a physical material, such as a stone in dried rice, a piece of bone in ground meat, a cherry or olive seed, or a piece of glass or metal in food resulting from a mishap during food manufacturing or processing, which may cause injury when biting down or if swallowed.

The term "RISK" can be defined as a function of the probability of an adverse effect and the severity of that effect, consequential to a hazard(s) in food. Risk is usually expressed as a ratio. It is an estimate of the chances you may have of being affected by the hazard in question, given that an appropriate assessment of the hazard was made and uncertainties were factored in.

Summary

1. The concept of risk is complex and multidimensional. It involves both objective aspects and individual perceptions, as well as societal values and preferences. In the common language, risk is associated with something undesirable, which may or may not happen, but gives reasons for concern. There are many different types of risks, but all are characterised by the real or perceived perspective of an uncertain but negative outcome. Various definitions are given, which put the accent on the possibility of an adverse outcome, its cause, its uncertainty or its consequence.
2. A risk is often coupled with an opportunity. Both risks and opportunities are related to uncertainty of outcomes, things that may or may not happen in future and possible but not sure negative or positive consequences. Since risk and opportunity are often linked, risk-taking is an important element of progress and innovation. Managing risks often requires striking a balance between risk and opportunity.
3. Although the psychological and context-dependent dimensions of risk must be considered in a comprehensive risk appraisal, the core of the concept of risk remains a potential adverse event taking place in the physical world causing real and measurable damage to something of value like health, safety, the environment etc., with a more than purely hypothetical probability to materialise.
4. In risk analysis, risk is conceived as an attribute of physical events with a probability different from zero to materialise. Nonetheless, for risk governance it is also necessary to take into account the concept of risk as a "social construct", a perception or an idea that is construed through cultural or social practice.
5. The technical definition of risk conceives risk as a function of the probability of an event and the severity and size of its adverse consequences. This definition focuses on the "hard facts" of risk and is central to the engineering practice, the EU regulation and the relevant international agreements. It has a quantitative connotation, but it cannot be always applied quantitatively. Sometimes, the definition is formulated as probability *times* severity and size of consequences. Nevertheless, it must be noted that although a high probability-low consequence event may, according to this formulation, be assigned the same level of risk of a low probability-high consequence one, two such situations may not mechanically be assumed to be equivalent in a risk management perspective.

6. Hazard and risk are given distinct meaning in risk analysis: hazard indicates the intrinsic capability of an agent, event, state of a system, course of action or behaviour, product, service, activity or technology to cause a negative impact on something that is considered valuable, like for example injury or harm to health, damage to ecosystems, damage to property or goods, financial loss etc. In order for the damage potential which is intrinsic in a hazard to materialise, several conditions must be met like, for example, a certain level of exposure, the presence of targets with a certain level of vulnerability etc. Confusion between hazard and risk may be a cause of inefficient regulation. Nevertheless there are cases where it is appropriate or even necessary to regulate on hazards.
7. Although the technical concept of risk was initially developed in an engineering perspective of risk quantification, it is often applied in a purely qualitative fashion, with different forms of grading or the use of qualitative and conventional terms like serious, negligible, acceptable, tolerable etc. Moreover, even in a systematic, science-based risk assessment process, a certain degree of subjectivity is inevitable, notably when assigning subjective probabilities to events or eliciting professional judgment.
8. The technical concept of risk is a simplification of a complex reality. Risk cannot be reduced to a single figure. Risks are not uniformly distributed and the same risk does not have the same meaning for all those exposed because the goods at stake may be valued differently. And the position towards a risk also depends on the associated benefits and opportunities, and on their distribution.
9. Effective risk governance requires consideration of the psychological and contextual dimension of risks. Perception of risk depends on its attributes and is enhanced if, for example, the risk is imposed, not controllable, uncertain, not familiar, unfairly distributed, dreadful etc.
10. Risk is a probabilistic concept. There is an intrinsic uncertainty in any risk. Such uncertainty can, in principle, be probabilistically quantified in an objective way. However, often additional uncertainties arise on the risk itself, which can only be qualitatively addressed or can only be quantified by assigning subjective probabilities. A distinction is therefore made between the decision under risk and the decision under uncertainty.

References

European Commission. (2015). *Better Regulation Guidelines*, SWD(2015)110 final.

European Union. (2002). *Regulation (EC) N° 178/2002 of the European Parliament and of the Council laying down the general principles and requirements of food law, establishing the European Food Safety Authority and laying down procedures in matters of food safety.*

European Union. (2006). *Regulation (EC) No 1907/2006 of the European Parliament and the Council concerning the Registration, Evaluation, Authorisation and Restriction of Chemicals.*

European Union. (2009). *Directive 2009/48/EC of the European Parliament and of the Council on the safety of toys.*

FAO (Food and Agricultural Organisation). (1997). *Risk analysis and code module 10, multilateral trade negotiations on agriculture-a resource manual.* Food Quality and Standards Service, Food and Nutrition Division.
IRGC (International Risk Governance Council). (2005). *Risk governance: Towards an integrative approach.*
ISO (International Standardisation Organisation). (2009). *Risk management-vocabulary,* Guide 73.
Kaplan, S., & Garrick, B. J. (1981). *On the quantitative definition of risk* (Risk analysis, Vol. 1, No 1).
Lowrance, W. W. (1976). *Of acceptable risk.* Los Altos: William Kaufmann.
Renn, O. (1992). Concepts of risk: A classification. In S. Krimsky & D. Golding (Eds.), *Social theories of risk* (pp. 53–79). Westport: Praeger.
SCCS (Scientific Committee on Consumer Safety). (2014). *Opinion on Acid Orange 7*, 18 June 2014, SCCS/1536/14, revision of 23 Sept 2014.
Schroder-Frechette, K. S. (1991). *Risk and rationality: philosophical foundations for populist reforms.* Berkeley: University of California Press.
Slovic P. (1987). Science 236 (17 April).
UK Royal Commission on Environmental Pollution. (1997). *Setting environmental standards*, 21st Report.
US NRC (Nuclear Regulatory Commission). (1975). *Reactor safety study-WASH 1400.*
WTO (World Trade Organisation). (1994). *Agreement on sanitary and phytosanitary agreement.*

Chapter 3
EU Risk Policy and Risk Analysis

Abstract We are daily faced with a great deal of risks that we manage on our own. Not all risks are regulated in our society. In fact, are usually regulated product and service risks that cannot be sufficiently controlled on the individual consumer-level, those that because of their nature or size go beyond the sphere of action of individuals and/or involve a public interest, like for example environmental or climate risks, as well as those related to externalities, situations where the risk-producer has no incentive to prevent or reduce the risk, or may even receive a benefit from imposing a risk on others. In advanced countries, we benefit from unprecedented, favourable conditions of life, including on health and safety, reflected in a dramatic increase of average lifetime, but new threats and risks have appeared, linked to technology, economic and population growth, which dwarf the ability of individuals to manage risks by themselves. Risks subject to regulatory action are often complex, uncertain or ambiguous. Risk managers therefore need to apply a structured approach. The concept of risk analysis, including the assessment, management and communication of risk, stays at the core of the EU and internationally recognised approach to risk. It is based on a technical definition of risk and involves a science-based approach to risk assessment and the functional separation between assessment and management. The risk analysis paradigm is integrated into the EU decision-making process, across many policy sectors, but important differences remain, which reflect historically different sectoral regulatory cultures and the specificities of the various legal frameworks.

3.1 Introducing Risk Policy

Risk is everywhere in everyday life. We are all "risk managers" when deciding where and when to cross the street, at what speed to drive in a busy highway (provided we respect any imposed speed limit), whether or not to apply a therapy recommended by our doctor or how to invest our savings. We all have an intuitive approach to risk. We are free to take or avoid, where possible, many of the risks that we face in our life. And we are responsible for most of the consequences. In fact, not all risks are or could or should be "regulated". It is part of our freedom and responsibility to make choices and protect ourselves against many of the great variety of risks that we encounter, or to decide to try our chance.

B. Delogu, *Risk Analysis and Governance in EU Policy Making and Regulation*,
DOI 10.1007/978-3-319-30822-7_3

Nevertheless, there are categories of risks that in organised societies are dealt with collectively, through public policies and regulatory activities. The borderline between public and individual responsibility for risks is variable in time, across jurisdictions and sectors, and is not necessarily consistent even within the same jurisdiction. It depends on many factors, like for example habits, preferences, history. In general, are regulated risks associated with products and services offered on the marketplace, to the extent that they cannot be sufficiently controlled by the consumer or the user; risks that because of their nature or size go beyond the sphere of action of individuals and/or involve a public interest, like for example environmental or climate risks; as well as those related to externalities, situations where the risk-producer has no incentive to prevent or reduce the risk, or may even receive a benefit from imposing a risk on others.

Nowadays, new types of risks deriving from technological development and the scale of human activities have dwarfed the ability of individuals to identify and manage risks by themselves. Public institutions are in particular called on to establish mandatory rules when there is a need to define conditions for activities potentially creating risks, like the placing on the market of certain products, the provision of services, the performance of professional activities, the introduction of new technologies, the execution of projects or the use of industrial processes. More recently, certain public policies have been established with the aim of influencing individual choices towards safer or healthier behaviour, like campaigns and measures to discourage smoking, promote healthy diets or safer driving modes.

Although an unprecedented reduction of the traditional risks and scourges has been achieved in our advanced societies, thanks to modern science, technology, medicine and social and technical organisation, new types of risks have emerged that require public policies and measures. The very nature, size and pervasiveness of certain new technologies, often acting in combination with the anthropic pressure caused by the enormously increased world population, create global risks that did not exist in the past. Man's activities for the first time can change irreversibly the climate and the environment on a global scale. Technology development is nowadays speedier than risk assessment and regulatory action. Globalisation and modern cross-border transport and travelling spread risks, notably risks of communicable diseases, across continents in a matter of days. Public expectations, perceptions and anxieties on new and emerging risks, often spread and amplified by the media, create political pressure in democratic societies that governments must address.

In all advanced societies, risk issues play therefore an important part in many public policy decisions. Decision makers in public institutions are often faced with risk issues when setting and implementing policies and rules pursuing broad objectives and protecting fundamental rights, like for example ensuring access to safe and healthy food, protecting or improving the quality of the environment or ensuring a high level of health protection. In view of such broad public policy objectives, policy makers are in most cases faced with issues that involve uncertainties and trade-offs. Risk issues rarely admit clear-cut, "black or white" assessment conclusions and management solutions. Easy choices, between clearly identified, safe-non safe alternatives, are quite rare. And a zero risk option is most often simply not available (some claim, as a matter of principle, that the zero risk does not even exist). Decision makers are in fact normally faced with several options, each characterised by its own potential for negative impacts, different probabilities for such impacts to materialise, different balance between costs and benefits, risk reduction potential versus ancillary risk appearance, as well as complex legal and societal implications. In order to manage the complexities of risk policies and measures, risk managers and political decision makers need the support of risk assessors who provide a science-based evaluation of the risks in presence. They need to appraise the risks, and to identify and assess the policy options in a multi-criteria perspective. Finally, they have the responsibility to take a political decision informed by both the scientific assessment and a comprehensive multi-criteria evaluation of all the dimensions of the risk issue.

EU risk policy is usually articulated on two different levels. The broad objectives and measures for the control of risks are generally established by general policy and legislative frameworks. In order to inform decisions on such general frameworks, risks are usually identified and assessed in a broad, generic and qualitative way. Examples of EU general regulatory[1] or policy frameworks are the EU Food Law, the REACH Regulation on chemicals, that we have already mentioned in Chap. 2, and the Action Plan on antimicrobial resistance (European Commission 2011). The first two are regulatory frameworks established by general legislation, while the Action Plan is a policy document.

On a second level, implementation measures address more specific and narrow issues. These measures take the form of specific norms, standards, limit values, restrictions, prohibitions or authorisations. Decisions on this level are normally supported by specific and detailed risk assessments, which are based, as far as possible, on quantitative methods, depending on the state of scientific knowledge and the availability of data.

3.2 The Concepts of Risk Analysis and Governance

We have introduced the concept of risk in Chap. 2. We have seen that risk is a both technical concept, which may be subject, to a greater or lesser extent, to scientific treatment and quantification, and a social and contextual construct. Many risks are seen by stakeholders and members of the public in a perspective that brings perceptions, preferences, values and comparisons together with what is known and understood by the layman about the factual and "scientific" aspects.

As previously noted, public perceptions and positions on risk-issues carry a heavy weight in risk management since the final political decisions that risk analysis

[1] In this work we do not make a strict terminological distinction between legislation and regulation since under the EU Treaties general legislation may be introduced by both "directives" and "regulations". Nevertheless, the different nature of the instruments referred to will be clear from the context of the discussion.

Table 3.1 Structure of risk analysis

RISK ASSESSMENT	RISK MANAGEMENT
• Hazard Identification • Hazard characterisation • Exposure assessment • Risk Characterisation	• Framing risk issues and preliminary definition of options • Defining scope and contents of risk assessment • Risk appraisal • Multi-criteria option assessment • Risk dialogue and communication
RISK COMMUNICATION	
• Two ways communication and dialogue across the assessment and decision-making chain and with the wide range of stakeholders and interested parties	

supports are in the end subject to democratic sanction in our institutional systems. Moreover, dealing effectively and efficiently with risks in democratic societies, characterised by complex social and institutional structures, requires the involvement, support and collaboration of a multitude of actors, over and above the own deliberations of the policy and regulatory authorities. A structured approach to risk policy needs therefore to integrate the technical paradigm of risk analysis with the broader concept of participatory risk governance.

3.2.1 The Concept of Risk Analysis

Components and Definition of Risk Analysis *Risk analysis* is generally understood to include the technical assessment of risk (Risk Assessment), the identification and evaluation of options leading to the establishment and implementation of policy and regulatory measures (Risk Management) and the two-ways communication between risk assessors and managers (Risk Communication) as shown in Table 3.1. Risk communication is nowadays given a wider meaning, covering also two-ways communication, consultation of, and dialogue with stakeholders, as well as information provided to the general public.

The concept of *risk governance* goes beyond the technical boundaries of risk analysis and includes a structured approach to the wider societal aspects of risk in a participatory perspective. In addition to the three steps of risk analysis, governance also includes organising public participation in the process, exploring stakeholders and public perceptions, concerns, preferences and values and framing the assessment and the appraisal of risk accordingly.

Risk Analysis Definitions and Guidelines The terms "risk", "risk assessment" and "risk analysis" are given different meanings and sometimes used in a confusing way in different policy and legal texts. Their understanding and interpretation may vary significantly across sectors and jurisdictions. Nevertheless, these terms have been somewhat codified in authoritative documents on the international, regional or national level. In particular, definitions and criteria for risk analysis have been provided in both the US and the EU.

The US National Research Council provided in 1983 definitions and guidance for the risk analysis paradigm in a seminal report "Risk Assessment in the Federal Government: Managing the Process", better known as the "Red Book" (US NRC 1983), without however using the expression "Risk Analysis". Although the Red Book approach focused on the assessment and management of the risks of chemicals, its concepts are transposable to most risk areas and have greatly influenced the approach to risk analysis across thematic borders and jurisdictions, including in the EU. The Red Book defines two stages in the decision making process related to risk control measures: risk assessment, defined as the characterisation of the potential adverse effects of human exposure to environmental hazards, and risk management, defined as the process of evaluating alternative regulatory actions and selecting among them.

According to the Red Book, risk assessment includes hazard identification, dose-response assessment, exposure assessment and risk characterisation. Risk management is a decision-making process and involves the integration of political, social, economic considerations and engineering information with risk-related information. The aim of risk management is to develop, analyse, and compare regulatory options. The final result is the selection of the appropriate regulatory option to deal with a potential hazard. The Red Book observes that the selection process necessarily requires the use of value judgements notably on the acceptability of risks and the proportionality of the control costs.

The Red Book recommends establishing and maintaining a clear conceptual distinction between the assessment of risks and the consideration of risk management alternatives. Such a distinction in practice requires the establishment of institutional settings and/or practical arrangements and procedures aimed at ensuring the operational separation of tasks and responsibilities between risk assessors and regulatory decision makers. Risk communication is not mentioned in the Red Book.

More recently, the US National Research Council re-visited the risk analysis paradigm in its reports of 2009 "Science and Decisions" (US NRC 2009). While confirming the basic principles and guidelines of the Red Book, the new report introduces important complements. It maintains the principle of separation between risk assessment and management, but the linear sequence assessment-management in the Red Book is replaced in the new report by a more interactive and comprehensive framework, where risk assessment is preceded by an initial formulation and scoping by risk managers of the problem to be addressed. Moreover, communication to, and involvement of stakeholder across the entire process is recommended as an integral part of the risk-based framework, while respecting the independence and integrity of the scientific advice.

The issue of separation between assessment and management was also discussed in a 1996 report of the US National Research Council: "Understanding Risk, Informing Decisions in a Democratic Society" (US NRC 1996). The report states in particular that the conceptual distinction between risk assessment (understanding) and risk management (action) remains useful notably for preserving the integrity of the scientific activity against any political pressure and maintaining the distinction between risk assessment and the cost considerations. Nevertheless, the report observes that a too rigid distinction does not help the improvement of decision-relevant understanding of risks because the analytical activities of risk assessment are not sufficient by themselves to provide the needed understanding.

3.2.2 Applications of the Risk Analysis Paradigm

Different Interpretations of Risk Analysis It is important to note that there are different interpretations of the risk analysis paradigm and the role of scientific risk assessment, and different approaches to risk policy. Tenants of the "technocratic" approach consider that dealing with risks is a matter for experts, scientific experts and technocrats, with a limited role for political institutions. Competence for both the assessment and the management of risks is then attributed to specialised agencies with regulatory powers, like the US Food and Drugs Agency-US FDA, responsible for food safety and pharmaceutical products. In such cases, the separation between risk assessment and management is purely functional, achieved through internal organisational measures aimed at ensuring a sufficient degree of independence of risk assessors from managers.

A stricter application of the risk analysis paradigm of the Red Book consists in the complete, structural separation between assessment and management. This is for instance the case of food safety in the EU, for which the EU political institutions have retained the regulatory responsibility[2] while risk assessment has been delegated to an external, independent body, the European Food Safety Authority. This approach is based on the assumption that risk assessment is a purely scientific activity and that risk analysis could be organised in a linear sequence: assessment, management, communication. As we will discuss further on, an approach isolating risk assessment from its policy context would not reflect the true nature of risk assessment, which is not pure science, and requires a sustained dialogue and interaction between assessors and managers.

A more realistic conception of risk analysis sees it as a circular, interactive process, integrated into the wider scheme of participatory risk governance, where the political institutions should define a risk assessment policy, framing the activities of assessors as regards all the normative, value-driven or subjective aspects of assessment. These are aspects on which EU practices have already made substantial progress since the post-BSE (Bovine Spongiform Encephalopathy) reform and with the Better Regulation working procedures, which will be discussed later in the book, within a trend towards a more mature, comprehensive and consistent approach.

[2] Certain regulatory responsibilities have been delegated by the European Parliament and the Council to the European Commission, which however is subject to the control of the two institutions for its delegated acts. Moreover, contrary to a common but erroneous opinion, the Commission is a political, not a bureaucratic institution, which is independent from the EU Member States but politically accountable to the European Parliament.

Whatever the structural solution adopted, following the various crises, like in particular that caused by the spread of the BSE in the 1990s, there has been from time to time and among contradictory developments a tendency of political decision makers to invoke "science" and technical expertise as a shield against responsibility for risks. It should be noted that the concept of "science-based" policy making is often criticised as functional to the attempt of de-facto attributing to scientists most of the responsibility for decisions on risk, presenting risk management decisions as characterised by the neutrality of science. In certain jurisdictions, the entire responsibility for dealing with risks has been transferred to technical bodies. On the EU level, that solution has not been chosen because of the constraints imposed by the Treaties[3] and the need to preserve the role of the Parliament and the Council in the complex, multi-layer and multi-national structure of the Union.[4]

3.3 The Risk Analysis Paradigm in the EU Risk Policy and Regulation

On the EU-level, the European Commission introduced the concept of risk analysis with its communication on "Consumer Health and Food Safety" of 30 April 1997 (European Commission 1997), where a new and comprehensive approach to consumer health protection was announced. Although the Communication specifically addressed food safety, its concepts were relevant in relation to the broader consumer health and safety regulatory approach.

The EU Definition of the Risk Analysis Concept The risk analysis paradigm in that communication included the three above-mentioned interrelated functions, risk assessment, risk management and risk communication, prescribing a clear separation of tasks and responsibilities between risk assessors and risk managers. Risk assessment was to be entrusted to independent scientific committees or bodies, established and acting in conformity to the principles of excellence, independence and transparency.

Risk analysis was defined in the Commission communication as

[3] See in particular the so-called Meroni doctrine, which arose from cases C-9/56 and C-10/56 (*Meroni v High Authority* [1957/1958] ECR 133), mentioning restrictive conditions for the delegation of powers, in particular that the delegation of power can only relate to clearly defined executive powers, the use of which must be entirely subject to the supervision of the delegating authority and that consequently it is not permissible to delegate broad discretionary powers that give to an agency a wide margin of appreciation. This makes it problematic for the EU delegating to an independent agency ample normative powers like those attributed to the US FDA.

[4] Nevertheless, the name of EFSA, which is designated as an "authority" (differently from other "agencies") although it has not been given regulatory powers, may perhaps be seen as a trace of some (frustrated) wish to "externalise" risk management together with risk assessment in the food safety area following the BSE crisis.

> A systematic procedure comprising the scientific evaluation of hazards and the probability of their emergence in a given context (risk assessment), the assessment of all measures making it possible to achieve an appropriate level of protection (risk management), and the exchange of information with all the parties concerned: decision-makers, inspectors, consumers and producers, in order to explain the reasons and to justify the management measures proposed (risk communication).

The Communication highlighted that decision-making responsibilities should however remain with the political institutions.

The Components of the EU Risk Analysis Paradigm In light of the definitions and criteria mentioned above, the core components and characteristics of the EU risk analysis paradigm are:

- The recognition of the role of independent scientific advice in informing decisions;
- The centrality of the concept of risk, characterised by both the probability and the severity of possible negative outcomes;
- The presence of the three interrelated but conceptually and operationally distinct functions: risk assessment, risk management and risk communication;
- A clear functional separation between risk assessment and risk management:
- The independence of the risk assessment function from policy makers and any vested interests;
- The consultative, non-binding and science-focused nature of risk assessment advice, which allows the integration of other legitimate considerations into risk management;
- A two ways communication with stakeholders on the risks, their assessment and the measures to control them. This last aspect has been much strengthened more recently as part of the EU "Better Regulation" approach that we discuss further on in this book, in a more participatory perspective compared with the wording of the above-mentioned Communication.

The Role of Science in the EU Risk Analysis Paradigm In establishing the risk analysis framework at the core of its policy and regulatory approach to consumer health, the Commission has recognised the role of scientific advice as the necessary basis for protection measures. Nevertheless, science is not the only factor determining and justifying such measures. In particular, the EU Food Law that we have already mentioned in Chap. 2, which takes on board and develops in legal terms the concepts of the Communication, states in its Article 6.3 that:

> Risk management shall take into account the results of risk assessment, and in particular, the opinions of the Authority referred to in Article 22, other factors legitimate to the matter under consideration and the precautionary principle where the conditions laid down in Article 7(1) are relevant, in order to achieve the general objectives of food law established in Article 5.

3.4 The Difficult Implementation of the Risk Analysis Paradigm

We will discuss further on in this book the qualifications, limitations and difficulties that must be addressed in order to ensure the utility and success of the risk analysis approach. The application of the risk analysis paradigm, where "science" plays a dominant part, is in fact constrained by several limitations that will be discussed in detail in this book. In particular:

Weak Public Confidence in Risk Science Policy makers, stakeholders and the public may attribute a limited value to scientific advice because:

- Risk "science" involves judgements and subjective assumptions, and is affected by uncertainties;
- Policy makers have difficulties in understanding the scientific aspects of complex risk problems and in resolving uncertainties and divergent scientific opinions;
- In general, public confidence in science, its independence, its ability to provide reliable advice have been compromised by divergences between scientists, cases of misconduct, misleading "scientific" statements about safety and the increasing dependence of science on business;
- While confidence in science as a guide for fundamental policy choices is limited, many think that values, in particular ethical considerations, precaution and societal preferences should instead be given a dominant role when dealing with the possibility of serious health and environmental damages, especially the global and irreversible ones.

Limited Availability of Independent Scientific Expertise, Knowledge and Data Access to independent high quality advice is becoming more difficult because:

- Knowledge and data on new technologies is mainly produced and detained by industry and business under proprietary conditions;
- There is little "independent" research and research centres and universities depend to an increasing extent on finance by industry;
- Technological progress and the related risk-issues develop at higher speed than the production of knowledge and data needed for risk assessment.

We will show in the next Chapters how these difficulties can be addressed by applying comprehensive governance approaches and risk-analysis best practices.

3.5 Risk Analysis and the EU Treaties

The Treaty on the European Union (TEU) and the Treaty on the Functioning of the European Union (TFEU) define the aims, principles, competences, policies and functioning of the European Union. The Treaties are supplemented by a Charter of

Fundamental Rights. Although there is no reference in the Treaties themselves to the concept of risk analysis, several of the policies established by the Treaties involve EU-level competences aimed or related to health, safety or environmental protection and involve EU-level risk management responsibilities.

EU Risk Management Competences Risk management responsibilities are indeed present in a wide range of EU policies. Some belong to EU exclusive competence areas like the conservation of marine biological resources and the common commercial policy. Others fall in the domain of shared competence, domains where both the EU and, under certain conditions, its Member States may act, notably the internal market, agriculture and fisheries, the environment, consumer protection, transport, energy or common safety concerns in public health matters. A few are in the domain where the EU competence is limited to co-ordinating and supporting national action: the protection and improvement of public health and civil protection.

In the case of the approximation of laws for the objective of establishing or ensuring the functioning of the internal market, the Treaty states that the Commission shall base its proposals concerning health, safety and environmental or consumer protection on a high level of protection, account taken in particular of any new development based on scientific facts (Art.114 TFEU, paragraph 3). The same Article of the Treaty requires any divergent national safeguard measures introduced or maintained after EU-level harmonisation to be justified on the basis of scientific evidence. The common transport policy includes a mandate for measures aimed to improve transport safety. The social policy includes measures aimed to set minimum requirements to protect workers' health and safety (Art. 153 TFEU). An entire article on public health (Art.168 TFEU) is aimed at human health promotion and protection. Protection of consumer health and safety is among the aims of consumer protection (Art.169 TFEU). Article 191 TFEU on the environment states that the Union policy on the environment shall aim at a high level of protection. Finally, in the area of civil protection, the Union shall aim in particular to support and complement Member States' action in risk prevention, preparedness and response (Art.196 TFEU).

Principles of the Treaties Some of the general provisions of the Treaties are at least indirectly relevant in view of framing EU risk analysis practice. Article 9 TFEU states that in defining and implementing its policies the Union shall take into account requirements linked to the promotion of a high level of human health. Article 11 TFEU says that environmental protection requirements must be integrated into the definition and implementation of the Union's policies and activities, in particular in order to promoting sustainable development. The already mentioned Article 191 TFEU on the environment states that the Union policy on the environment shall be based, among others, on the precautionary principle, and that in preparing its policy the Union shall take account in particular of available scientific and technical data and the potential benefits and costs of action or lack of action. Similar principles are stated by the Charter (Articles 35 and 37 respectively), which has the same legal value of the Treaties.

In conclusion, the EU has very extensive competences across many and diverse policy areas for action related to various types of risks (health, safety, environmental, occupational etc.). The Treaties do not refer to a risk-based approach in relation to the EU "protection" objectives on health, safety and the environment, and the concept or components of risk analysis are not mentioned. The EU approach to risk control in the areas of health, safety and the environment is established in a sector-based perspective by the secondary EU legislation and in guidance documents, and is also guided by the relevant case-law of the Court of Justice.

3.6 Risk Analysis Concepts in Some Regulatory Instruments

A variety of EU instruments and actions implement the principles and objectives of the Treaties on health, safety and environmental protection. This section examines whether, where and how risk analysis concepts appear in some of the EU legal instruments most relevant to health, safety and environmental protection. Some of the legal instruments selected are aimed at law harmonisation in various sectors (technical harmonisation for the internal market). The main objective in such cases is to harmonise the rules applicable to the placing on the EU market of the products considered. Some other Directives or Regulations considered here are aimed at setting minimum rules on environmental protection and the related human health protection aspects.

The sample of EU legal instruments selected here is not intended to be representative of the entire range of the relevant EU legislation. Nevertheless, it covers a significant range of policy areas and regulatory models. The sample considered in this assessment includes legislation on food (general food law, novel food, food-contact materials), cosmetics, pharmaceutical products, chemicals, the environment (air quality, integrated pollution control), medical devices, toys, and electrical products (low voltage).

3.6.1 Food Safety

General Food Law The already mentioned Regulation (EC) No 178/2002 of the European Parliament and the Council establishes the EU General Food Law and the European Food Safety Authority (EFSA). The concept of risk analysis is at the core of this Regulation.

Recital 16 of the Regulation states that measures adopted by the Member States and the Community should generally be based on risk analysis, except where this is not appropriate to the circumstances or the nature of the measure. Recital 17 affirms that the three interconnected components of risk analysis-risk assessment, risk management and risk communication-provide a systematic methodology for the determination of effective, proportionate and targeted measures or action to protect health. Recital 18 affirms that risk assessment should be undertaken in an independent, objective and transparent manner, on the basis of the available scientific information and data. Recital 19 recognises the role of legitimate factors other than scientific risk assessment that may need to be taken into account for a risk management decision and recitals 20 and 21 refer to the precautionary principle.

Risk, risk analysis and risk assessment are specifically defined in Art. 3. Art.6 sets out provisions for risk analysis, which is referred to as the basis of EU measures and action on food safety "except where this is not appropriate to the circumstances or the nature of the measure". Art.7 introduces the precautionary principle and defines its relations with risk assessment. Finally, Chapter III establishes and regulates the European Food Safety Authority. It provides for the separation between assessment and management and attributes to the Authority the responsibility for risk assessment. The regulation ensures the independence of the Agency from policy makers. Independence, competence and transparency are the basic principles across the various provisions concerning the functioning of the Authority.

The food law is the first piece of EU legislation that has expressly and formally integrated the risk analysis paradigm. It is of particular importance for the level of detail and clarity of its definitions and provisions, its internal consistency, but also its wide scope and implications for both the EU legislation and the national provisions on food and feed. Nevertheless, the legal implications of the general food law for the EU sectorial legislation is a complex matter. In spite of the intention to complete an adaptation of all relevant sectorial instruments to the principles of the general food law by 2007, not all the texts have been revised. Moreover, it is unclear how far the general food law limits the discretion of the EU legislator in introducing restrictions and conditions, like for example authorization requirements, not supported by specific risk assessments.

Novel Food Regulation (EC) No 258/97 of the European Parliament and the Council (European Union 1997) on novel food applies (until 1 January 2018) "to the placing on the market within the Community of foods and food ingredients which have not hitherto been used for human consumption to a significant degree within the Community". It states that novel food and its ingredients may only be placed on the market if they do not "present a danger for consumers", among other conditions. It provides for a notification procedure leading to an initial assessment by a national "food assessment body", followed, if necessary, by an additional assessment on the EU level by the Scientific Committee for Food (now replaced by EFSA) aimed to verify compliance with the "absence of danger" criterion, as part of a centralized authorisation procedure.

The regulation does not specifically refer to the concept of risk. It does not include criteria or procedures for the assessment on novel food. In practice, the role attributed to the European Food Safety Authority by the General Food Law ensures that the assessment is conducted in accordance with the horizontal risk-based approach established by the General Food Law.

A first attempt to put the Novel Food Regulation in line with the principle of the General Food Law failed in 2011 on disagreement between the EU Institutions concerning cloning in food production. A new proposal for revision was presented by the Commission in 2013 and a new Regulation has finally been adopted by the European Parliament and the Council in November 2015 and will apply from 1 January 2018 (European Union 2015). It establishes a centralised authorisation regime managed by the Commission. In order to be authorised a novel food must not (among other conditions). "…on the basis of the scientific evidence available, pose a safety risk to human health". Before updating the list of authorized novel food, the Commission will request EFSA to give its opinion as to whether the update is liable to have an effect on human health. In assessing novel foods EFSA is required in particular to:

> "…where appropriate, consider whether:
>
> (a) the novel food concerned is as safe as food from a comparable food category already placed on the market within the Union;
> (b) the composition of the novel food and the conditions of its use do not pose a safety risk to human health in the Union."

The preamble of the proposed regulation includes references to the scientific basis of the authorisation regime, to the role of EFSA and to the application of the precautionary principle in case of uncertainties.

Food-Contact Materials Regulation (EC) No 1935/2004 on food contact materials (European Union 2004-1) requires that such materials "must not transfer their constituents to food in quantities which could endanger human health".

Specific measures are established for groups of materials or articles. EFSA must be consulted on provisions liable to affect public health. Substances not yet included in a positive list following a specific measure are subject to an authorisation procedure involving a "safety assessment" by EFSA.

The regulation states that where a draft specific measure is not in accordance with the opinion of the Authority, the Commission shall provide without delay an explanation for the reasons of the divergence.

3.6.2 Cosmetics

A new Cosmetics Regulation, Regulation (EC) No 1223/2009 of the European Parliament and the Council, is applicable since July 2013 (European Union 2009a). The previously applicable directive, Council Directive 76/768/EEC, under which many of the currently available products have been placed on the market, stated that a cosmetic product "must not cause damage to human health". To that aim, the person responsible for the placing on the market must keep available for control information, including an assessment of the safety for human health of the product, with the general toxicological profile of the ingredients, chemical structure and level of exposure. The assessment was to be done by a "qualified person holding a diploma in pharmacy, dermatology or similar". Member States were empowered to prohibit a complying product if it represented "a hazard to health". Specific implementation measures were taken by the Commission on ingredients prohibited or allowed under condition and/or for a limited period, "after consultation of the Scientific Committee for Cosmetic Products and Non-Food Products Intended for Consumers" (this Committee, established by a Commission Decision, has changed name several times). The Directive did not include details on the mandate of the Committee or the criteria for establishing its opinions.

The new regulation introduces provisions closer to the risk analysis paradigm. Nevertheless, the new text does not fully integrate the generally accepted risk analysis terminology and concepts. The regulation states that a cosmetic product "must be safe". The person responsible for marketing the product must ensure that the product has undergone a safety assessment and that a safety report is set up. Detailed guidelines for establishing a safety report are provided in an Annex. However, those guidelines do not refer to the concept of risk and are not aligned with the commonly accepted risk assessment approach. Also, criteria for assessing possible risks and establishing the level of safety are missing. Only a few of the concepts that intervene in a risk assessment are mentioned.

The Regulation requires that the Commission Scientific Committee on Consumer Safety (SCCS) be consulted by the Commission on implementation measures. This requirement applies in particular when amending the lists of ingredients prohibited or restricted, where there is "a potential risk to human health, including when there is insufficient data". The SCCS must also be consulted on the safety of nano-materials and the safety for use of carcinogenic, mutagenic or repro-toxic (CMR) substances.

Interestingly, the Regulation includes some hazard-based provisions aimed at the ban on certain categories of carcinogenic, mutagenic and repro-toxic substances from use in cosmetic products, on consideration of the very direct exposure of users of such products and the absence of adequate benefits balancing the possible risks.

Although the Regulation entrusts the SCCS with an important consultative role, it does not set out specific provisions for its mandate, composition, working procedures and the applicable risk assessment criteria. Those are established by the Commission in a separate Decision.

3.6.3 Pharmaceutical Products

The EU legislation on pharmaceutical products for human and veterinary use includes several legal instruments, in particular the Council Regulation (EEC) No 726/2004 (European Union 2004-2) and Directive 2001/83/EC (European Union

2001), supported by related instruments. The legislation sets out rules for the authorisation, placing on the market and the post-marketing surveillance (pharmacovigilance) of medicinal products. The authorisation is subject to a procedure intended to demonstrate the quality, safety and efficacy of the products. The legislation sets out requirements on the tests to be performed, as well as the information and data to be provided, which include in particular pharmaco-toxicological and clinical test-data.

A distinct set of requirements, established by a specific regulation (European Union 2014), which has recently substantially modified and up-dated previous provisions, apply to the conditions for clinical trials.

Under a centralised procedure, the authorisation of certain pharmaceutical products is decided upon by the Commission assisted by a Standing Committee of representative of the Member States, account taken of the opinion of the European Medicines Agency (EMA). EMA delivers opinions on the granting of marketing authorisations in light of the evaluation of the risk-benefit balance of the product, which includes consideration of quality, safety and efficacy of the product.

Although the Commission and the Member States retain the power to authorise medicinal products, the Agency is involved in risk/benefit management decisions by delivering substantive, even if formally non-binding, opinions on the authorisation itself. Therefore, the Agency's task goes beyond "informing" the management decision. This is confirmed by the requirement on the Commission to motivate any divergence from the recommendations of the Agency.

In the case of pharmaceutical products, the risk analysis paradigm is adapted to the specific nature of the products considered, which requires integrated assessment of risks and benefits.

3.6.4 Chemicals

The placing on the market of chemicals in the EU is strictly regulated by REACH, the already mentioned Regulation (EC) No 1907/2006 of the European Parliament and the Council on registration, evaluation, authorisation and restrictions of chemicals. REACH aims at ensuring a high level of protection for human health and the environment. This regulation sets out comprehensive provisions related to the registration, evaluation, authorisation and restriction of chemicals. It establishes a European Chemicals Agency (ECHA) including in particular a Risk Assessment Committee (RAC) and a Committee for Socio-Economic Analysis (SEAC).

A fundamental aim of REACH is to ensure, in a staged approach, the availability of information and data on all the chemicals placed on the EU market above the threshold of 1 tonne per year. REACH integrates the risk-based approach into all levels of its requirements, processes and procedures. Art.44 of REACH states that prioritisation of substances for evaluation must be risk-based. Although the identification of substances of high concern, which are subject to an authorisation procedure, is based on hazard characteristics, the authorisation itself depends on consideration of risks and socio-economic benefits (the latter considered when risks cannot be "adequately controlled"). The process is informed by opinions of the Committee for Risk Assessment and, where appropriate, the Committee for Socio-Economic Analysis. In the case of chemicals posing "an unacceptable risk to human health or the environment", REACH establishes a procedure for adopting marketing restrictions that includes the consultation of RAC and SEAC.

Risk assessment and risk management (authorisation, restriction) responsibilities are functionally separated. The Commission and the Member States, assisted by SEAC, retain management tasks while RAC is responsible for risk assessment. However, the tasks attributed to SEAC fall well within the risk management area.

It must be noted that REACH includes provisions for the assessment of substances aligned with the usual risk assessment paradigm (hazard identification and characterisation, exposure assessment, risk characterization). Moreover, comprehensive guidance for both health and environmental risk assessment of chemicals is developed through implementation projects.

3.6.5 Air Quality

Directive 2008/50/EC of the European Parliament and the Council (European Union 2008) establishes provisions for the assessment, monitoring and management of air quality in the EU. The directive introduces a simple approach aimed at ensuring that concentrations of certain air pollutants are adequately assessed and monitored and air quality plans established, with the overall aim to keep the exposure of the public within safe levels and to preserve air quality as far as feasible. There is no specific provision for a risk-based approach to quality standard setting in this directive although health protection objectives are the basis for the limit and target values of pollutants concentrations and duration of exposure.

The Directive states that such values are "fixed on the basis of scientific knowledge".

3.6.6 Integrated Pollution Control

A comprehensive up-dated approach to integrated pollution prevention and control has been established with Directive 2010/75/EC (European Union 2010) of the European Parliament and the Council on industrial emissions, recasting, updating and completing several older legal instruments. The Directive applies to a wide range of industrial activities. It establishes conditions for the operation of such activities aimed at preventing, reducing or eliminating pollution as far as possible. It regulates emissions into all the environmental media. The directive adopts a technology-based approach. Emission limits and other applicable requirements are based on the best available techniques (BAT). BAT are defined as "the most effective and advanced stage in the development of activities and their methods of operation which indicates the practical suitability of particular techniques for providing the basis for emission limit values and other permit conditions designed to prevent and, where that is not practicable, to reduce emissions and the impact on the environment as a whole". In turn, available techniques mean "those developed on a scale which allows implementation in the relevant industrial sector, under economically and technically viable conditions, taking into consideration the costs and advantages, whether or not the techniques are used or produced inside the Member State in question, as long as they are accessible to the operator". Best "means the most effective in achieving a high general level of protection of the environment as a whole".

This directive is based on a "state of art" approach. It does not involve case by case assessment of risks for health and the environment. Therefore, the concept of risk is not specifically relevant under the approach of the directive as emissions must in any case be reduced as much as the BAT does allow.

3.6.7 Medical Devices

The objective of this "new approach" directive, Council Directive 93/42/EEC (European Union 1993) amended several times and currently being again revised, and complemented by directive on active implantable and in-vitro diagnostic medical devices, is to harmonise the requirements related to the safety and clinical quality of medical devices. To that aim, it establishes essential requirements, conformity assessment procedures and market surveillance provisions. The essential requirements of the directive take into account a risk/benefit balance: "the devices must be designed and manufactured in such a way that, when used under the conditions and for the purposes intended, they will not compromise the clinical condition or the safety of patients, or the safety and health of users or, where applicable, other persons, provided that any risks which may be associated with their intended use constitute acceptable risks when weighed against the benefits to the patient and are compatible with a high level of protection of health and safety". A number of specific requirements are then established in qualitative terms, notably that the risks must be eliminated or reduced as far as possible, protection measures taken against the remaining risks and information provided where residual risks remain in spite of the protection measures. Across the text, it is stated that risks must be "eliminated", "reduced", "minimised", "as far as possible" or "to the lowest possible level".

Depending on the "class" of the device, a third-party (notified body) assessment is required. Notified bodies are entrusted the task of verifying compliance with the essential requirements, including the risk/benefit balance, and issuing conformity certificates. The Directive does not specify the methods and criteria for conformity assessment. However, according to the "new approach", implementation of the essential requirements is often ensured through harmonised standards that set the reference for conformity assessment. In practice, the depth of the risk assessment performed by notified bodies depends on whether harmonised standards are applied and on the degree of detail of the applied standards. In case of devices containing medicinal substances, the notified body must seek the opinion of a competent authority or EMA on the clinical benefit/risk profile of the substance.

In conclusion, the directive on medical devices is based on a risk/benefit approach. Its implementation involves the assessment of risks by manufacturers and notified bodies, with an important role for the standardisation bodies. Market intervention of the surveillance authorities are based on risk-benefit considerations. Nevertheless, the Directive does not include specific requirements on risk assessment and the methods and criteria to be applied in that respect.

3.6.8 *Toys*

EU legislation on toys was revised and re-cast in 2009 by Directive 2009/48/EC of the European Parliament and the Council (European Union 2009b). Like many other directives on technical harmonisation (in particular, Directive 2006/95 on electrical products referred to below), Directive 2009/48/EC is a "new approach" directive setting essential safety requirements and conformity assessment procedures and referring to harmonised standards for detailed safety specifications. Nevertheless, differently from other directives of the same category, the new Toys directive has specifically integrated a risk-based approach. It provides specific definitions for "hazard" and "risk". Risk terminology is used across the text.

The directive establishes safety requirements related to various types of risk. Nevertheless, based on precautionary considerations, the safety requirements for certain chemical substances of high concern in toys are based on hazards instead of risks. Manufacturers are required to perform and document a safety assessment addressing both hazards and exposure before placing a toy on the EU market. Conformity assessment of toys involves a risk assessment covering all the relevant safety aspects. Finally, re-evaluation of risks posed by certain chemicals in toys is entrusted to an independent Scientific Committee.

3.6.9 *Electrical Products*

Directive 2006/95/EC of the European Parliament and the Council (European Union 2006) regulates the placing on the EU market of electrical products. The Directive sets out the safety requirements applicable to equipment designed for use with a voltage range between 50 and 1,000 V (alternating current) and 75 and 1,500 V (direct current), with a few exceptions. The Directive includes essential safety requirements and conformity assessment criteria, the latter referred to various categories of safety standards. The current "Low Voltage" Directive is the revised (2006) version of an older (1973) Directive. The concepts and terminology of the old directive have been maintained and the concepts of risk and risk assessment do not appear in the text.

This Directive does not follow a risk-based approach. Products must be designed in a way ensuring protection against hazards. Various categories of hazards are mentioned in general terms (danger of physical injuries or other harm which might be caused by direct or indirect contact; temperatures, arcs, radiation etc.). No specific criteria are established on the way in which “persons and domestic animals” must be “adequately protected”. The qualitative, hazard-based safety requirements of the directive are the basis for the establishment by the European Standardization bodies of the relevant harmonized standards that are in practice the reference for the design of products and the assessment of conformity. Harmonised standards define the design characteristics ensuring that exposure to the relevant hazards is reduced in order to ensure “adequate” protection.

3.6.10 Risk Analysis in EU Regulation

An Uneven Approach to Risk Concepts in the EU Legislation Protection of health, safety and the environment are at the core of a great number of EU legislative and implementation measures. The general criteria and the approach at the basis of the relevant instruments and measures have great practical relevance and significant legal, economic and trade implications. The simple, sample-based examination of the relevant provisions of the Treaty and secondary legislation presented above shows that there is no single, easily recognisable risk-paradigm informing the relevant instruments across the various sectors.

Although the EU general policy orientation for regulation on health, safety and the environment is meant to be aimed at a risk-based approach, there is no uniform conceptual and terminological approach in the relevant instruments across sectors. For example, the concept of risk is at the basis of the food safety legislation and is also well recognisable in the legislation on pharmaceutical products, chemicals and toys, but it is absent in the directives dealing with the safety of electrical products and integrated pollution control. Terms like risk and hazard are in some cases used without definition and distinction, and sometimes both appear to be used interchangeably in the same text.

The objective of the assessment prescribed by the legislation is sometimes defined as the verification of the safety of a product, but in other cases is the establishment of the level of the risk posed by it. It is not clear whether these different formulations by the legislator are intentional or derive from some confusion on the concepts of hazard, risk and safety.

In the EU legislation, the risk assessment paradigm is mentioned and comprehensively defined only in the food law, where it was shaped by the major food crisis caused by the spreading of BSE (Bovine Spongiform Encephalopathy). That paradigm is closely translated into legislative provisions and institutional arrangements for food safety, notably in the role attributed to EFSA as well as the provisions on the structure and functioning of the Authority. Elements of the risk analysis paradigm are found in other pieces of legislation, but in most cases without an expressly formulated, recognisable, coherent and complete framework of definitions and provisions. An example is provided by the legislation on pharmaceutical products, where the separation between risk management and risk assessment is not clearly established even if the formal decision-making power of the Commission is preserved.

Most of the relevant legislative instruments include provisions on consumer information or public access to information (safety of products, environmental quality etc.). Nevertheless, communication as part of the risk assessment/management process is in general not mentioned.

The EU Scientific Advice Structure The EU structure providing scientific and technical advice and support in the area of health, safety and the environment covers a wide range of domains and includes several bodies as shown in Table 3.2 (SCENIHR and SCHER have been recently-April 2016-merged into a single committee, the SCHEER):

However, the bodies in question are based and work on significantly dissimilar principles even in the case of tasks of similar nature. That is particularly visible in relation to independence and separation between assessment and management tasks. The mandate of EFSA and the Commission Scientific Committees is strictly limited to risk assessment (EFSA has been given in addition certain tasks of technical support to the Commission that are compatible with its role as independent risk assessor). The rules for selecting and appointing the experts in charge of the scientific advice as well as the rules of procedures of those bodies are intended to ensure a complete separation and independence from all sort of influences, including political and national interests. In the case of EMA and ECHA, the national authorities intervene in the designation of the experts, who in turn maintain a link with the respective national bodies. Moreover, EMA has the task of pronouncing on risk management, e.g. recommending the authorisation of pharmaceutical products. ECHA goes beyond the assessment of risks and provides socio-economic assessment of measures, a task belonging to risk management in the risk analysis paradigm. It should also be noted that while risk analysis places the concept of risk at the core of the regulatory approach, there are cases where EU legislation has established provisions based on consideration of hazards. We will discuss the conditions and justifications for a hazard-based approach to health protection in Chap. 5.

In conclusion, as it is the case in other jurisdictions, although the concept of risk analysis or some of its elements are present in the EU legislation addressing health, safety and environmental issues, the risk analysis paradigm does not consistently constitute, from a formal and legal point of view, an overarching conceptual framework for the relevant regulatory activities across the sectors. That is due to historical reasons, the establishment of sectorial legislation over a long period of time, the arising of specific regulatory models in different areas, which are difficult to change even when a regulation is updated or recast. However, the risk-based approach has gained a dominant status in the EU in practice, across all the sectors and independently

Table 3.2 The EU scientific advice structure

RA body	Area of competence
EFSA	Food and feed safety, animal health and welfare, plant health
EMA	Safety/effectiveness medicines human use; safety/effectiveness medicinal products for veterinary use, pharmaco-vigilance
ECHA	Registration, evaluation and registration of chemicals (REACH)
ECDC	Communicable disease, surveillance, preparedness and response
EEA	Air, water, soils pollution, climate change, natural resources and bio-diversity
SCENIHR	Emerging or newly identified health risks
SCHER	Risks related to toxicity and eco-toxicity of chemical, bio-chemical and biological compounds
SCCS	Health risks of non-food consumer products
SCOEL	Occupational exposure to chemicals

of the diverse formulations in the legal instruments, notably as part of the better regulation policy with its impact assessment provisions that require a technical assessment of risks whenever appropriate. We say more on that in Chap. 5 on risk management and in Chap. 7 on regulatory policy and governance.

Summary

1. We are daily faced with a variety of risks that we have to manage on our own. Responsibility for taking, avoiding or facing and managing many types of risks is in the first instance an individual responsibility and part of our freedom. Not all risks are regulated. And even when there are rules, like for example in the case of road safety, our behaviour and choice matter.
2. In organised societies are dealt with collectively, through public policies and regulatory activities or other instruments, product and service risks that cannot be sufficiently controlled on the consumer level, those that because of their nature or size go beyond the sphere of action of individuals and/or involve a public interest, like for example environmental or climate risks, as well as those related to externalities, situations where the risk-producer has no incentive to prevent or reduce the risk, or may even receive a benefit from imposing a risk on others.
3. New types of risks deriving from modern technological development, the scale of human activities and the unprecedented size of world population have dwarfed the ability of individuals to identify and manage risks by themselves. And in our democratic systems, risk issues as seen through public perception and expectations, play an important part in the political debate.
4. Risk issues are often complex, uncertain or ambiguous. In order to manage such difficult issues, political decision makers need the systematic support of scientific advisors, structured communication and dialogue approaches as well as mechanism to involve a wide range of actors in the risk governance process. They have the responsibility to take a political decision informed by both the scientific assessment and a comprehensive multi-criteria appraisal of all the dimensions of the risk issue.

5. At the core of the EU risk policy stays the concept of risk analysis. Risk analysis includes the technical assessment of risk (Risk Assessment), the identification and evaluation of options leading to the establishment and implementation of measures (Risk Management) and the two-ways communication between risk assessors and managers (Risk Communication). While risk analysis focuses on the technical dimensions of risk, the effective governance of controversial risks requires also consideration of the wider societal perspective, through a participatory process.
6. Although the more traditional view of risk analysis used to conceive risk assessment as its purely scientific pillar, strictly distinct from risk management, a more realistic conception sees risk analysis as a circular, interactive process, integrated into the wider scheme of participatory risk governance, with a need for sustained dialogue between assessors, managers, policy makers and stakeholders.
7. The risk analysis paradigm takes the science-based assessment of risk as the central piece of the process. However, scientific risk assessment is not without difficulties. Policy makers, stakeholders and the public may attribute a limited value to scientific advice notably because risk "science" involves judgements and subjective assumptions, and is affected by uncertainties. Moreover access to independent high quality and timely available advice is becoming more difficult because of the growing influence of industry on science, the rapid technological development that supersedes scientific research on risks and the difficulty to access proprietary data and information.
8. On the EU level, the risk analysis paradigm has been integrated into consumer health and food safety policy in the aftermath of the BSE food crisis at the end of the 1990ies.
9. The EU has extended competences for regulating health, safety and environmental risks across a wide range of policies and sectors. Nevertheless, there is no reference to a risk based approach or to risk analysis in the EU Treaties themselves. Definitions and provisions related to the risk analysis paradigm are found in some EU regulatory instruments, notably the food safety legislation, and the risk based approach is included in the regulatory guidelines (Better Regulation) of the European Commission ensuring that it is taken into account as appropriate on the preparatory stage of regulatory proposals.
10. A dedicated, comprehensive scientific risk assessment advice structure has been put in place by the EU, ensuring support across a wide range of sectors. It includes specialised advisory agencies and scientific committees.
11. The examination of a sample of the relevant EU legislation shows that there is no uniform approach in the legal texts to risk, risk assessment and risk analysis in general. The various instruments have been conceived in different periods, within different legal frameworks and also reflect historically different regulatory cultures. Nevertheless, a gradual convergence is taking place in practice toward a more homogeneous risk-based approach, notably within the framework of the Better Regulation guidelines that promote horizontal integration and consistency.

References

European Commission. (1997). *Consumer health and food safety*, COM (97)1997.

European Commission. (2011). *Action plan against the rising threats from antimicrobial resistance*, Communication, COM 748, 2011.

European Union. (1993). *Council Directive 93/42/EEC concerning medical devices.*

European Union. (1997). *Regulation No 258/97 of the European Parliament and the Council concerning novel foods and novel food ingredients.*

European Union. (2001). *Directive 2001/83/EC of the European Parliament and the Council on the Community code relating to medicinal products for human use.*

European Union. (2004). *Regulation No 1935/2004 of the European Parliament and the Council on materials and articles intended to come into contact with food and repealing Directives 80/590/EEC and 89/109/EEC*, 2004-1.

European Union. (2004). *Regulation No 726/2004 of the European Parliament and the Council laying down Community procedures for the authorisation and supervision of medicinal products for human and veterinary use and establishing a European Medicines Agency*, 2004-2.

European Union. (2006). *Directive 2006/95/EC of the European Parliament and the Council on the harmonisation of the laws of Member States relating to electrical equipment designed for use within certain voltage limits.*

European Union. (2008). *Directive 2008/50/EC of the European Parliament and the Council on ambient air quality and cleaner air for Europe.*

European Union. (2009a). *Directive 2009/48/EC of the European Parliament and the Council on the safety of toys.*

European Union. (2009b). *Regulation No 1223/2009 of the European Parliament and the Council on cosmetic products.*

European Union. (2010). *Directive 2010/75/EU of the European Parliament and the Council on industrial emissions (integrated pollution prevention and control).*

European Union. (2014). *Regulation No 536/2014 of the European Parliament and the Council on clinical trials on medicinal products for human use, and repealing Directive 2001/20/EC.*

European Union. (2015). *Regulation (EU) 2015/2283 of the European Parliament and of the Council on novel foods.*

US NRC (National Research Council). (1983). *Risk assessment in the Federal Government: managing the process.* Washington, DC: The National Academies Press.

US NRC (National Research Council). (1996). *Understanding risks, informing decisions in a democratic society.* Washington, DC: The National Academies Press.

US NRC (National Research Council). (2009). *Science and decisions.* Washington, DC: The National Academies Press.

Chapter 4
Risks and Risk Assessment

Abstract Specific risk assessment methods have been developed for different types of risks. Nevertheless, they present some common features. This Chapter focusses on the assessment of chemical, biological, ecological and radiation risks, the safety of food and feed derived from genetically modifies plants, and on the safety assessment of nuclear power plant. The aim is to introduce the basic methodological issues and problems of risk assessment through the example provided by the methods applied in a range of sectors which are of particular interest for the EU regulatory action. The basic paradigm of risk assessment includes the identification and characterisation of the hazards, the assessment of the exposure to the relevant stressors, the characterisation of the risk integrating the information on the two dimensions of risk: the hazard and the exposure. Risk assessment methods try to quantify the risk, but in many cases only a qualitative expression is possible. Risk assessment is often affected by uncertainty and variability. Probabilistic methods have been developed to address and represent uncertainty. Several aspects of risk assessment are characterised by a certain degree of subjectivity, notably when evidence and data are weighed, defaults are selected, assumptions are made based on professional judgment and expert opinion elicitation or subjective probabilities are assigned. Transparency on such aspects is key for the credibility of the assessment process. Close collaboration between assessors and managers is needed in order to frame the assessment within a *risk assessment policy* setting criteria and guidance for the value-laden or subjective aspects of the process as well as for the functioning of the interface between assessment and management.

4.1 Introduction

This is the longest and more technical Chapter of the book. It does not need to be read in its entirety at once, but the reader is advised to read at least the sections on the general aspects of risk assessment and that on the assessment of chemical risks before proceeding to the remaining Chapters. The reason is that the discussion on the general aspects and the detailed description of the assessment methods for chemical risks, and on the related issues, provide the necessary background for the discussion on risk management and governance, and on risk assessment best practices. The considerations on uncertainties, precaution, the role of evidence in

B. Delogu, *Risk Analysis and Governance in EU Policy Making and Regulation*,
DOI 10.1007/978-3-319-30822-7_4

decisions, the need for dialogue between assessors and managers, the relation between science and policy, and many other aspects treated in the remaining of the book would sound quite abstract in the absence of a good understanding of the nature, potential and limits of risk assessment. In that respect, the assessment methods for chemicals provide a useful reference paradigm for a better understanding of the discussion on all those issues.

Although quite technical and detailed, this Chapter is not a short risk assessment "manual". Nobody would be able to perform risk assessments in any of the fields considered here just after studying its contents. Risk assessment requires specific training in a number of complex scientific disciplines and can only be performed by experienced scientists. And there is a wealth of dedicated textbooks, manuals and courses available to those who wish to specialise in the various relevant aspects, from toxicology, epidemiology, pharmacology, ecotoxicology to radioprotection or engineering safety assessment. The aim of this Chapter is instead to introduce the non-specialist reader to the basic concepts, issues and problems of risk assessment and to provide information on the broad lines of the specific approaches in some important risk sectors.

Detailed information is provided on the methodological aspects of the toxicological assessment of chemicals, while for ecological risk assessment and the assessment of microbiological risks the presentation focuses only on the similarities and specificities compared to the case of chemicals. Finally, health risks of radiation and the safety of nuclear power plants, and the relevant assessment methods, both subjects of considerable technical complexity, are presented here only in broad, descriptive terms, providing some background information that may help the understanding of related physical, engineering and methodological aspects.

4.2 Risk Assessment: General Aspects

This Chapter focuses on a few risk-issues related to health, safety and environment protection, among those that are most relevant within the framework of the EU regulatory competence. It introduces the methods for the identification and assessment of the relevant risks. This introduction to risk assessment does not intend to, and could not be exhaustive. There are a variety of methods adapted to the specific nature of the many different types of risks associated with technologies, products, technological or natural systems. Nevertheless, the broad lines of the technical risk assessment approach are similar across sectors and risk-issues. In order to provide a concrete perspective on the risk assessment paradigm, the discussion of chemical risks and their assessment is developed here in some detail, both because chemical risk assessment plays a particularly important role in the EU risk-policies across many sectors and because it provides a useful paradigm of the basic risk assessment concepts applicable, with the necessary adaptations, to many other sectors.

The discussion on the microbiological, ecological and radiation risks and risk assessment methods illustrates the similarities and specificities across sectors, while the brief introduction to the safety risks of nuclear plants is of particular interest for explaining the concepts and methods of probabilistic risk assessment. Nuclear plant safety also provides examples of major unforeseen accidents caused by the difficulty of identifying, characterising and anticipating all the relevant hazards and their probability of occurrence in complex systems and situations.

As anticipated in the introduction, the aim of the Chapter is to provide the knowledge and understanding of the concepts, terminology, methods, problems and limits of risk assessment to the extent necessary for the proper and critical interpretation of the scientific risk assessment advice provided by experts and specialised scientific committees.

The basic paradigm of risk assessment consists of a few conceptual elements applicable across many specific sectors. The starting point is the identification and characterisation of the potential causes of risks, the "hazards". The hazards are determined by those properties, situations, events that have an intrinsic potential to cause damage to something of value under certain circumstances. In some cases it is possible to reach a good, deterministic knowledge of such causes, but often significant uncertainties appear about the reality, relevance, severity, size and characteristics of a hazard.

The further step is the study of the conditions under which the damage resulting from the intrinsic characteristics of the hazard in presence may materialise. In the case of chemical health risks this step includes studying the level, duration, frequency of the human population exposure and the characteristics of the exposed population, notably its vulnerability. Uncertainty assessment plays again an important role at this stage.

The final step is the characterisation of the risk resulting from the conjunction of the intrinsic hazard and the conditions under which the hazard is deployed in a specific situation. Risk assessment aims at quantifying risks whenever possible, but often only a qualitative assessment is feasible.

When discussing the concept of risk, we have highlighted its multidimensional nature. Risk assessment generally focusses on the technical dimension of risk. In spite of its limits, technical risk assessment is a key tool to inform risk management decisions, without prejudice to the more comprehensive analysis and appraisal that risk managers need to perform, bringing together all the other policy-relevant dimensions of risk. At the same time, risk managers should ensure that the scope and issues for a risk assessment are properly and comprehensively defined in order to serve the needs of policy makers and respond to the concerns and interests of all relevant actors and interested stakeholders.

We have also noticed in our discussion on the concept of risk that risk is about uncertainty and lack of knowledge. Risk assessment brings in the tools of science in the attempt of providing a solid ground for decisions. However, there are limits to what science can deliver. Risk assessment necessarily includes conventional approaches, models, extrapolations, assumptions, which fill gaps in the available data bases and in the scientific knowledge. It is in the end the task of risk managers to resolve the remaining uncertainties and unknowns. Therefore, although risk assessment is a task for scientists, account taken of its complexities, uncertainties and theoretical and practical limitations involved, it is important that policy makers, who assume the responsibility of the final decisions, get a critical insight into its nature and methods.

In light of the above-mentioned considerations, elaborating on and completing a broad definition given by the US National Research Council in 1994 (US NRC 1994), we conceive in this book (technical) risk assessment as a systematic, science-based approach to organising and analysing knowledge, data and information for potentially hazardous activities, technologies, products, substances that might pose risks under specified circumstances. Risk assessment also includes the study of the conditions under which the hazard can materialise and the integration of the information about hazard with that on the specific circumstances, notably the exposure to the stress factor, leading to an actual adverse effect.

4.3 Introduction to the Assessment of Heath Risks of Chemicals

We introduce in this section the methods for the assessment of health risks associated with the exposure to chemicals. Across this Chapter, the term "chemicals" indicates chemical substances that may consist of elemental molecules or compounds made of two or more chemical elements.

After the description of the assessment methods, the reader will find in this section a few paragraphs on some more general aspects and issues of chemical risk assessment, notably the difficulty for the non-specialist to interpret risk assessment advice, the need for a common understanding and "language" for facilitating the dialogue between assessors and decision makers, the importance of uncertainty in risk assessment, the new perspectives for methodological improvements, the definition of the scope for risk assessment and the conventional assumptions at the basis of risk quantification. Several of these issues, which are here introduced in relation to the specific case of chemicals, are in fact also relevant for risk assessment in general and could therefore be read as part of a general introduction to risk assessment. In order to complete the background to our discussion on the risk assessment of chemicals, the next section provide some background on the importance of the chemical industrial sector in the EU economy, its key role in technological development and innovation, but also its risk-potential and the need for a regulatory framework balancing risks and benefits and ensuring longer term sustainability. Finally, Sect. 4.3.2 shows how the EU is at the front line of regulatory development for chemicals, notably with its comprehensive REACH Regulation on registration, authorisation and restriction of chemicals.

4.3.1 *Benefits and Risks of Modern Chemistry*

Chemicals are everywhere around us, in the products that we use and in the built and natural environment in which we live. Modern chemistry and its products contribute in a fundamental way to advances across the most diverse technological sectors, and to continuous improvements in our living conditions. Life-saving medicines, pain-killers, new lighter, stronger materials for a wide range of products and applications, detergents that allow ensuring high hygiene standards, fertilisers and pesticides that increase agriculture productivity and contribute to reduce malnutrition and hunger, are all achievements of modern chemistry.

Chemistry is really at the core of progress and wellbeing, and contributes to the strategies to face the challenges of the future. However, production, use and disposal of chemicals, and of products containing chemicals, may involve significant risks for human health and the environment. Such risks depend on both the intrinsic characteristics of the chemicals and the level of human and environmental exposure associated with the various life-cycle stages of the applications or products considered. Experience shows many examples of disastrous consequences, often unexpected, of the use of certain dangerous chemicals, from thalidomide to asbestos or CFCs, as well as examples of serious accidents in the chemical industry, like Bhopal and Seveso.

As many as 143,000 chemical substances, currently marketed in the EU, were pre-registered under the EU-REACH registration scheme by the 1st December 2008 deadline. That figure provides an illustration of the importance of chemicals in our modern life and the size of their presence in our environment. Moreover, according to the European Confederation of Federations of Chemical Industries (CEFIC), chemical industries in the EU have contributed in 2013 to the EU economy for 558 billion EUR, employed 1.2 million workers and produced 20 % of world chemicals.

4.3.2 *EU Legislative and Regulatory Decisions on Chemicals*

Safe use of chemicals, in a lifecycle perspective, is therefore of the utmost importance for protecting health and the environment, for fostering sustainable innovation and for ensuring the long-term development and sustainability of such a key technological and economic sector. For those reasons, the EU has established the most comprehensive regulatory framework in the world on chemical safety. In addition to the general regulatory framework established by REACH, the already mentioned EU Regulation on Registration, Authorisation and Evaluation of Chemicals, all the most relevant applications of chemicals are specifically regulated. Examples are: medicines, fertilisers, plant protection products, food additives and food-contact materials, detergents, cosmetics.

Implementation of this comprehensive regulatory framework is supported by scientific bodies, independent from the Institutions, providing EU decision makers with risk assessment advice. In implementing the objectives of the Treaties and of the secondary legislation, EU decision makers are in fact very often confronted with the need, on the basis of risk-considerations, to decide on bans, restrictions or authorizations of certain chemical substances or preparations in order to ensure the protection of health, safety and the environment. In order to identify the most suitable option for action, in view of a political decision, risk managers need to weigh risks, costs and benefits.

Some of the relevant decisions taken at EU-level have a wide and general scope, while others relate to restrictions or marketing conditions for particular products or applications. REACH is an important example of legislation with a very wide scope, addressing a broad range of objectives: health, safety and environmental protection as well as the harmonisation, for the objectives of the internal market, of the conditions for the placing on the market and use of chemicals in the EU.

Implementation measures based on REACH usually target specific categories of chemicals or individual substances. Similarly, general safety requirements are established for cosmetics by the Cosmetics Regulation, which are supplemented and updated through more specific decisions.

Under the current EU regulatory practice, independent scientific advice on the potential risks associated with the chemicals considered contributes to inform regulatory decisions in a structured and systematic way. Scientific risk assessment advice is complemented by socio-economic and legal assessment, ensuring consideration of other relevant aspects like benefits, costs, alternatives, societal preferences and legal, industrial, social and trade implications. As noted in the previous Chapters, in the case of REACH, a risk assessment committee (RAC) works in parallel to a Committee for Socio-Economic Analysis (SEAC) in order to provide a comprehensive assessment of all the relevant aspects. Moreover, the European Commission now applies to all major regulatory measures a comprehensive impact assessment procedure that brings together information on all the relevant aspects in a systematic and integrated manner (We discuss Better Regulation and the Impact Assessment in the EU regulatory practice in Chaps. 5 and 7).

Health risk assessment of chemicals for policy and regulatory needs is intended to inform decision making: the overall objective is to provide decision makers with information on the existence, nature, level and extent of risks for human health that may be associated with certain chemicals used in a wide variety of activities, applications or products. Health risk assessment is sometimes also performed for needs other than regulatory ones, in different contexts like, for example, insurance for setting prime-level, or research for defining funding priorities. In such cases, the methodology may need to be adapted to the specific objectives and needs pursued.

4.4 Approach and Methods for the Health Risk Assessment of Chemicals

4.4.1 The Stepwise Structure of Health Risk Assessment of Chemicals

As said before, the methodology for health risk assessment of chemicals has an interest beyond its direct field of application and provides a reference paradigm for risk assessment in many fields. It plays an important role in supporting the EU and international regulatory practice. The next titles in this section introduce in some detail the basic methodological concepts of health risk assessment of chemicals, covering its two main aspects:

- the study of the intrinsic characteristics of the chemical considered in relation to its potential for adverse effects in humans (its hazards), and
- the evaluation of the health relevance of a chemical in a particular application or product by assessing the exposure and by combining the data on hazard and exposure in order to provide information on the resulting risk.

Health risk assessment is organised as a sequence including the four basic steps of Fig. 4.1.

Definitions, issues and methods of each step are presented in the following paragraphs.

4.4.2 Step 1 of Risk Assessment: Hazard Identification

Hazard identification is the search for and determination of any intrinsic property of the chemical studied having the potential to cause an adverse health effect in humans exposed. There are many types of possible health hazards of chemicals and a given chemical may show more than a single hazard.

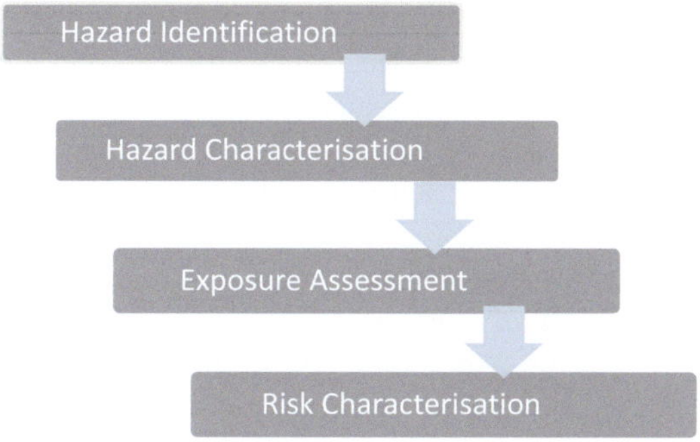

Fig. 4.1 Steps of health risk assessment

Chemicals May Show Different Types of Hazards, Related to Different Modes of Exposure In the identification of hazards it is taken account of the mode of exposure, distinguishing between inhalation, ingestion or dermal contact. The mode of exposure has an influence on the delivery of the chemical to the organism. A chemical may in fact show a certain hazardous property only in relation to a given mode of exposure (for instance when it is inhaled).

Hazards May Be Identified by Different Methods, but Most Data Currently Used for Risk Assessment of Chemicals Come from Animal Tests Hazards of chemicals may be identified through *in vivo* or *in vitro* tests (respectively tests on animals or on isolated biological components like tissues or cells), epidemiological studies (studies aimed at identifying an adverse health effect in a population or group exposed to a chemical) and/or study of the chemical structure of the substance and its relation with a possible adverse effect (*in-silico* studies). Results from one type of tests constitute a "line of evidence". Convergent results from more than one line of evidence reinforce an indication of the presence or absence of danger.

Although most data currently are still produced by tests on animals, in vitro and in-silico studies, together with the "omics" tools provided by the latest developments of molecular biology and informatics, are taking an increasing role, also in view of the particularly important objective of reducing or replacing animal testing (on "omics", see paragraph 4.5.3).

In-Vitro, In-Silico and Epidemiological Studies In-vivo studies are discussed in paragraph 4.4.3. The nature of in-vitro, in-silico and epidemiological studies deserves some clarification.

In vitro studies are aimed at examining the effects of a substance by testing its action on isolated elements of an organism like cells or biological molecules. These studies allow a substantial simplification by eliminating the influence of the extremely complex functional systems represented by the organisms from which the individual components are extracted. Nevertheless, such a simplification is also the limit of in vitro studies as the results of in vitro tests must then be extrapolated to the biology of an entire organism. The interpretation, in terms of health damage potential, of in vitro results showing an effect at cell-level is sometimes very difficult in the absence of complementary evidence.

In-silico studies are based on consideration of the molecular structure of chemicals and the link between such structure and the resulting biological activity. They are based on the principle of similarity that states that compounds with similar structures may be expected to show similar biological properties. In-silico studies make use of quantitative structure-activity relationships or read-across data for similar compounds. In pharmacological research, in-silico studies involve building and running computer models simulating cellular mechanisms and conducting computer-based tests of a given substance. These studies require the introduction of high volumes of data. They are often part of integrated testing strategies aimed at minimising animal tests and are useful for screening purposes and for identifying priorities for in vitro or in vivo tests.

Epidemiological studies are aimed at comparing groups of people which differ for the exposure to a risk factor or for the incidence of a pathology in order to identify or verify a causal relationship between a factor and a health effect. There are many different types of epidemiological studies. Frequently referred to in risk assessment reports are:

- *Cohort studies*: a group of people representing a sample of a population exposed to the risk factor considered, for instance to a dangerous substance, is monitored over a period of time and both the exposure and the incidence of the pathologies studied are recorded. The data resulting from the monitoring are analysed and compared to data from the observation of a control group of non-exposed people in order to establish whether a statistically relevant causal correlation can be established between the level of exposure and the incidence of the pathology. Cohort studies may be prospective, following the sample group into the future and recording the current exposure, or retrospective, when past exposure conditions are considered and the follow-up is forward in time.
- *Case-control studies*: the exposure history of members of a group of people showing the studied pathology is compared to that of a control group not affected in order to establish whether the different health status can be attributed to difference in the exposure to the risk factor studied.
- *Cross-sectional studies*: both the current health status and the exposure level of a group are assessed in parallel. This study provides a picture of the population sample considered at a given time in relation to both exposure and incidence of an effect. It does not allow establishing a cause-effect link for effects that can appear in the longer term.
- *Ecological studies*: the aim is to compare the incidence of the pathologies considered in a given geographical area to that observed in another area, for example to compare the incidence of cancer in an area with high environmental concentration of pollutants with that of a non-polluted area

Various indicators are used to express the results of epidemiological studies, notably:

- *Odds ratio* (*O/R*): the ratio between the incidence of a disease in a case-control study for the exposed group and in the non-exposed group. For example, if the incidence in the exposed group is twice that in the non-exposed one, the O/R is 2.
- *Standardised Mortality Ratio* (*SMR*): the relative death occurrence in the exposed to the non-exposed group. An SMR of 1.50 means that the mortality is 50 % higher in the exposed group.
- *Relative Risk* (*RR*): same as SMR, but referred to the incidence of a disease instead than death.

Epidemiological studies have the advantage of providing information based on direct observations on humans, without a need for extrapolating from animal models. Nevertheless, the design and the interpretation of such studies are often challenging, a long time may be needed to produce results from cohort studies and they are in general costly and exposed to bias and confounding factors. Moreover, they do not allow the ex-ante identification of health risks. This explains why most risk assessment of chemicals is based on data from animal tests.

Human studies involving intentional exposure to potentially hazardous chemicals aimed at producing health damage pose obvious ethical problems and would in most cases be illegal. Nevertheless, some useful human data on the delivery of chemicals to organ and tissues or their mechanisms of action are produced by preclinical and clinical studies in the pharmaceutical research sector.

Hazard Identification Is the Pre-condition for a Rational Approach to Risk A chemical may show hazard characteristics without posing health risks when used in a particular application that does not involve human exposure. On the contrary, it is obvious that there could not be risks in the absence of hazards. Nevertheless, public debates on risks related to purely hypothetical hazards are frequent, notably when new products or technologies are introduced. The identification of a hazard or at least the determination based on relevant evidence and scientific considerations of the plausibility of a given type of hazard is the necessary condition for a rational approach to risk assessment and management.

A Biological Effect Is Not Necessarily an Indicator of an Adverse Health Effect An important and often challenging task in hazard identification is to discriminate between adverse health effects and biological effects with no health relevance. That objective is complicated by the need to extrapolate from the effects observed in animal tests, or in the case of *in-vitro* tests in isolated tissues and cells, to possible adverse health effects in humans. Certain biological effects observed in animal tests may in fact be health-relevant in animals but not in human organisms.

Interpretation of Data from Animal Tests Requires the Application of Statistical Analysis in Order to Detect Significant Effects As said before, most of the available data on health hazards of chemicals come from test on animals. The objective of such tests is to detect those adverse effects on animals that are considered to be relevant for human health. In practice, an effect is detected when a statistically significant difference is observed for a biological indicator or for the incidence of an adverse health outcome or of death, between the groups of test-animals exposed to different doses of the chemical and a reference group of non-exposed animals of the same species and type and observed under identical conditions. The size of the groups must be sufficient to ensure the statistical significance of the observed results. The sensitivity of the test, that is its ability to detect an effect, increases with the size of the test groups. However, there are principle and practical limits to the number of animals that can be used.

The duration of the study must be adapted to the type of potential effect explored, which may vary from acute, short-term toxicity, to chronic, long-term effects. Best-practice test methods for identifying health hazards of chemicals are codified and internationally recognised.

In Application of the EU Legislation, Chemicals Are Classified According to Their Hazards EU legislation provides for the classification and labelling of chemical substances and preparations according to their hazards, including in particular hazards related to human health. The EU system is put in line with the internationally agreed Global Harmonised System (GHS). Main types of human health hazards, some of them subdivided in various categories, are:

- *Acute Toxicity*, defined on the basis of the lethal dose causing the death of 50 % of test-animals exposed for oral, dermal or inhalation exposure.
- *Skin Corrosion*, defined as the irreversible damage to the skin following the application of a test substance.
- *Skin Irritation*, defined as the production of reversible damage to the skin following the application of a test substance
- *Serious Eye Damage*, defined as the production of tissue damage in the eye, or serious physical decay of vision, following application of a test substance to the front surface of the eye, which is not fully reversible within 21 days of application.
- *Eye Irritation*, defined as changes in the eye following the application of a test substance to the front surface of the eye, fully reversible within 21 days of application.
- *Respiratory Sensitization* defined as hypersensitivity of the airways following inhalation of the substance.
- *Skin Sensitization*, defined as an allergic response following skin contact.
- *Germ Cell Mutagenicity*, defined as an increased occurrence of mutations in populations of cells and/or organisms.
- *Carcinogenicity* defined as the induction of cancer or the increase of its incidence.
- *Reproductive Toxicity* defined as adverse effects on sexual function and fertility in adult males and females, or developmental toxicity in offspring.
- *Specific Target Organ Toxicity* includes all significant reversible, irreversible, immediate or delayed health effects that can impair function for a target organ, distinguishing between single and repeated exposure effects.
- *Aspiration Hazard* covers severe acute effects such as chemical, different degrees of pulmonary injury or death following the entry of a liquid or solid into the trachea and lower respiratory system, directly through the oral or nasal cavity or indirectly from vomiting

4.4.3 Step 2 of Risk Assessment: Hazard Characterisation

The general assumption in health risk assessment of chemicals is that the appearance and extent of an effect caused by exposure to a chemical depend on the dose: "*The dose makes the poison*" (Dosis facit venenum-Paracelsus, sixteenth century). Hazard characterisation is the study of the quantitative relation between the dose and the effect.

The Concept of Dose *The dose* is the quantity of a chemical delivered in a given period of time to an organism through one or more routes such as ingestion, inhalation or dermal contact.

Effects generally depend on the weight of the individual exposed. Therefore, the dose is normally expressed in milligrams or micrograms per kilogram of body weight and per day or week).

The "*external*" *dose* refers to the quantity entering the organism while the "*internal*" *dose*, refers to the quantity delivered to the critical organ or tissue.

Chronic effects are related to doses repeated over a very long period or the entire lifetime. Acute effects are related to high doses delivered over a short period of time.

Different approaches are followed to characterise hazards depending on the type of effects. There are typically two different situations, depending on the existence of a threshold for the effects observed, which lead to two different approaches for the assessment of risks.

Non-threshold Effect: Quantitative Risk Assessment Certain categories of hazardous chemicals are considered to be able to cause an adverse effect at any dose. The extent of the effect depends on the dose. Such effects are usually referred to as "non-threshold effects". In such cases, the aim of risk assessment is to quantify the risk for the exposed population that is associated with a certain level of exposure (for example the expected additional incidence of cancer in a group exposed), or the level of exposure below which the risk is considered to be "acceptable" (for example the exposure below which the increase in lifetime incidence of cancer is less than 1 in a million).

The typical non-threshold effect is the genotoxic carcinogenicity, a property of chemicals that may cause cancer through gene damage. Genotoxicity refers to the ability of a substance to cause damage of the DNA resulting in mutations or cancer. The DNA (deoxyribonucleic acid) is the double-stranded molecule composed by four bases (substances conventionally indicated with the letters A, T, C and G from the initials of their denomination) that allows the registration and reproduction of the genetic code presiding to the development and functioning of an organisms (notably by encoding RNA, a molecule transporting the messages contained in the genetic code, and proteins). The way in which the bases are arranged determines the "message" registered in the genome. The human genome, that has been completely deciphered, includes three billion pairs of bases. The bases are organised in sections, the genes, which in the human genome are between 20,000 and 25,000. The configuration of the bases in the genes is of fundamental importance for the development and functioning on the organism. Damage to DNA, if not repaired by its internal mechanisms, may have very serious health consequences. For genotoxic substances, the assumption is that even a single molecule in a single hit can cause the DNA damage in a cell that initiates a degenerative process leading to cancer.

Threshold Effects: Safety Assurance Other adverse effects called "threshold effects" are instead considered to manifest themselves only above a certain dose level. For such effects, the objective of risk assessment is in general to establish the level of exposure below which there is no adverse effect for people exposed, including, in general, sensitive individuals.

The theoretical assumption for threshold chemicals is that if exposure remains below a certain dose-level, the organism is able to resist against the potential damage through homeostatic and cyto-protective processes. Homeostasis is the ability of a biological system, a cell, an organ or an entire organism to maintain or re-establish its internal conditions by feedback regulation and control processes, in order to stabilize health and functioning, in presence of changing external conditions or the action of stress factors. The term cyto-protective indicates a factor or mechanism that protects cells from noxious chemicals or other stress factors.

The threshold assumption is also considered valid for non-genotoxic carcinogenicity. Genotoxic carcinogenicity is instead treated in all cases as non-threshold, as explained above, because the current state of knowledge does not allow taking account of possible gene repair mechanisms in the characterisation of such hazard.

Animal Studies Animal studies are performed in accordance with international guidelines. They involve extrapolations and use of safety/uncertainty factors, which are conventional or specific depending on data available. Hazard characterisation aims at establishing a dose-effect curve or another type of mathematical model allowing the calculation of the risk level or of a safe level of exposure.

In animal studies, researchers observe the effects on various "endpoints" of a chemical which is delivered with a particular mode of exposure and at different dose-levels to various groups of animals. "Endpoint" is defined as an effect observed in a toxicity study. The effects are selected based on their biological significance and potential health concern. An example is liver damage. The endpoints may be expressed as a continuous variable, like the average size of weight increase of an organ, or a discrete "quantal" variable like the number of animals in which a certain lesion has appeared. A quantal effect is an effect which is expressed as occurring or not occurring like death or appearance of a tumour.

The observed effects are compared with the outcome in a comparable reference group of non-exposed animals. As noted before for hazard identification, in order to conclude to a causal relationship between the different dose-levels and the variation of the effects observed, the size of the study groups must be large enough to detect statistically significant differences between the test groups and compared to the reference group. Using several groups exposed to different doses allows determining the dose dependency of the effect considered in the range of exposure applied in a toxicity study.

Types of Animal Tests, Guidelines and Good Practice There are standardized methods for testing a wide range of toxic effects on animals. The most relevant effects studied are:

- *Acute Toxicity*: Tests to determine toxic effects from a single or brief exposure; short observation period (2 weeks)
- *Sub chronic Toxicity*: Tests to determine toxic effects from repeated exposures from several weeks to several months (typical observation period: 90 days)
- *Chronic Toxicity*: Tests to determine the toxic effects from exposure covering a substantial part of lifetime (typical observation time: 12–24 months)
- *Carcinogenicity*: Tests to determine carcinogenic effects (typical observation periods depend on the test animals used, notably 18–24 months for mice and 24–30 months for rats)
- *Reproductive Toxicity*: Tests to determine effects on gonadal function, conception, birth, growth and development of offspring (Observation period: test substance given to parental animals prior to mating, during pregnancy and through weaning of offspring; substance then given to selected offspring during their growth into adulthood, mating and production of a second generation, until the second generation is 21 days old).
- *Developmental Toxicity*: Tests to determine embryotoxicity effects and birth defects (Observation period: offspring sacrificed and examined day prior to expected birth for teratogenic effects; offspring observed for growth retardation and abnormal function through infancy and examined for teratogenic effects).
- *Dermal Toxicity*: Tests to determine the irritation and inflammation of the skin (Short observation period; there are two tests, with and without sensitization)
- *Ocular Toxicity*: Tests to determine effects on eyes, irritation or corrosion (Observation period: 72 h)
- *Genetic Toxicity*: Tests to determine gene mutation, chromosome changes, and DNA activity; there is a variety of tests on insects, rodents microorganisms and plants.

Animal tests confer a considerable added value to toxicological studies because they allow to control exposure of the test groups, as well as the environmental conditions, to a high degree of precision and uniformity, make it possible to evaluate different types of toxic effects, and, differently from in-vitro tests, to explore and understand the toxicity mechanism in a complete organism.

The number of animals to be used is determined by the type of effects studied and the need to ensure statistical reliability and sensitivity. Under current regulatory requirements on the animal welfare the number of animals used must be reduced to the minimum. The species that are used depend on the type of toxicity test that are performed. Rodents and rabbits are the most commonly used laboratory species due to their availability, low costs in breeding and housing, and success in producing reliable results. Generally, tests are designed adopting for the exposure the route (inhalation, oral, or dermal) corresponding to human exposure; the age of test animals should be representative of that of humans; dose levels are chosen with a view to determine thresholds and dose-response relationship for the effects studied.

There are however several limitations to this type of studies, with heavy implications for the method itself and for the meaning and interpretation of the results.

First, tests are conducted with very high doses, normally well above the doses to which people are exposed, because at lower doses the number of test animals needed to detect effects in a statistically significant way would reach unmanageable levels. As a consequence, for quantitative risk assessment dose-effect data from animal tests must be extrapolated downward, by use of mathematical models, by as much as three or four orders of magnitude, i.e. by a factor 1000 or 10,000, to the much lower dose levels which are representative of human exposure. The dose-effect relationships in the range of the low doses that result from such extrapolation depend heavily on the model used. Very different conclusions may therefore be reached for the hazard characterization of a chemical substance depending on the way animal data are extrapolated from high to low dose levels. This intrinsic limitation of animal model method is a source of significant uncertainty.

Second, results of tests on animals are extrapolated to humans. Moreover, there is a need to account for the variability of sensitivity across the exposed human population, depending on factors like age, state of health, sex, ethnicity, etc. In order to account for *interspecies* and *inter-individual* differences, the reference point data (for instance, the highest dose at which no statistically significant effect is observed) resulting from animal tests are divided by safety or uncertainty factors. Such factors may be based on specific evidence when sufficient knowledge is available on the rate of internal delivery to organs or tissues, the mechanisms of action of the chemical studied, and if there is adequate data on the population exposed. Unfortunately, in most cases such knowledge and data are lacking. Therefore, the safety/uncertainty factors are in most cases based on conventional assumptions.

Use of safety factors is well established practice, but there is still uncertainty and some disagreement in the scientific community whether the used factors are adequately representative of interspecies and inter-individual variability. The factors used are considered over-conservative by some authors, while other authors claim that, in certain cases, they might not be protective enough.

Characterization of Mixtures No harmonized approach is available yet for the characterization of mixtures. Animal studies are in general performed exposing test animals to a single chemical. They do not provide information on the potential effects of exposure to mixtures of chemicals. As we will explain later on, there are no harmonised methods yet to assess such combined effects.

Threshold and Non-threshold Effects The different approach for threshold and non-threshold effects is based on theoretical considerations more than empirical evidence. It should be noted that, although different approaches are conventionally used in practice for threshold and non-threshold effects, there is no final agreement on the existence of effects that show no threshold. In general, both for threshold and non-threshold effects, it would not be possible in practice to directly observe the effects of chemicals at low doses in animal studies as the sensitivity of tests with a manageable number of test-animals would not be sufficient to detect statistically significant differences between the test-group and the reference-group. Therefore, in the absence of solid empirical evidence, the different approaches followed for the two types of effects are justified by theoretical considerations, as explained before.

The conceptual differences in the approaches applied to the characterisation of hazards for threshold and non-threshold effects lead to two different methods on which some further considerations are necessary:

4.4.3.1 Method of Hazard Characterization for Threshold Effects of Chemicals

The "Safe" Dose Concept For threshold effects, the characterisation method is aimed at identifying a dose reference-point from which a "safe" exposure level is calculated. As we have explained before, in the case of non-genotoxic effects, effects that show a threshold, the objective of hazard characterization is to determine a dose level which is expected to be "safe" for humans, in other terms the dose level below which the likelihood of adverse health effects in humans is considered to be "very unlikely."

For that aim, different groups of test animals are exposed to decreasing doses of the chemical to be assessed in order to determine the lowest dose at which an adverse effect is still statistically detected (LOAEL: Lowest Observed Adverse Effect Level) or the highest dose at which no adverse effect is observed (NOAEL: No Observed Adverse Effect Level).

In practice, it is not feasible to determine in a precise way the threshold for the activity of the chemical in animals. The NOAEL, which is the reference value most frequently used in practice, is a surrogate for such a threshold. It is by definition lower than the "real" threshold value. How close a NOAEL is to the real activity threshold depends on the spacing of the test doses in the animal study (the more narrow is the difference between the doses used, the closer the NOAEL will be to the real value), on the sensitivity of the test and the size of the groups of test animals (both increase the ability to detect "real" change in response to the change in the dose). Large spacing of doses increases the probability to select a too conservative value for the NOAEL (a dose level that is significantly lower than that at which no adverse effect would be observed in a test with narrower distance between the doses used), while poor sensitivity of the test or too small groups work in the opposite direction (effect not detected at a dose where in reality there is one).

The NOAEL should not be misunderstood for a dose-level at which there is no adverse effect at all as its denomination could suggest. The NOAEL is not necessarily a "no adverse effect" dose, although it is sometimes interpreted as such.

No-observed-effect levels determined in animal studies must be reduced in order to account for interspecies differences. Differences between individuals must also be considered. To that aim uncertainty/safety factors are used. In order to transfer to humans the NOAEL determined in animal studies the NOAEL is usually divided by so-called safety or uncertainty factors. A factor of 10 is applied to account for the interspecies variability. A further factor of 10 is applied to account for the variability of sensitivity across the exposed human population. These are conventional default factors that may be replaced by different factors when specific relevant knowledge and data are available on the action mechanism of the chemical and the characteristics and sensitivity of the population exposed. As a consequence of the application of the two uncertainty factors, the "safe" level for humans is in general the NOAEL divided by 100, that is a dose one hundred times lower than the highest dose associated with no observed adverse effect in animals.

The uncertainty or safety factors used for taking account of interspecies and intraspecies variability are intended to cover the influence of respectively species and individual differences related to both toxicokinetics and toxicodynamics.

Toxicokinetics is the process of the uptake of potentially toxic substances by the body (how and how much of a substance enter the body), their possible transformation by biological processes (biotransformation) into other substances (the "metabolites"), the distribution of the substances and their metabolites in the tissues, and the elimination of the substances and their metabolites from the body. Toxicodynamic is the process of interaction of potentially toxic substances with target sites in the body and the subsequent reaction leading to an adverse effect.

In order to facilitate the substitution of the conventional values of ten for the two factors, when more specific data is available, it has been suggested that the 10 factor for interspecies differences should be decomposed into two sub-factors, 4 for toxicokinetics and 2.5 for toxicodynamics ($4 \times 2.5 = 10$). For the inter-individual differences, the factor may also be decomposed into the product of two sub factors as the product $3.2 \times 3.2 = 10.24$, slightly greater than 10, respectively representing the toxicokinetic and toxicodynamic aspects. In this way specific knowledge on a particular aspect of the variability could be more easily incorporated into the factor by replacing the conventional sub-factor with the specific one.

As previously highlighted, the standard safety factors are somewhat arbitrary since in most cases there is no evidence supporting the assumption that interspecies and inter-individual variability would be represented by a factor of 10 independently of the type of chemical and the specific effect considered. Nevertheless, those factors are widely used and have been considered to be adequately protective and even in most cases conservative. Some authors have even considered that those default factors also cover possible combined effects of mixtures, even if they were originally not intended for that aim. On the contrary, other authors have expressed doubts about the adequacy of the factors to ensure sufficient protection, notably in the case of mixtures, referring in particular to higher interspecies variability observed in certain tests on different animal species.

4.4.3.2 An Example of NOAEL Use

In order to be more concrete on the meaning and use of NOAELs, we anticipate a bit here on the discussion about risk characterisation by providing an example of the use of a NOAEL in practice that the reader may wish to study in order to check how

Table 4.1 Critical endpoint and NOAEL for di-methyl phthalate

Name	Maximum perfume concentration (mg/kg)	NOAEL mg/kg/day Critical effect
Di-methyl phthalate (DMP)	2982	600 Maternal toxicity

the concepts discussed above apply to real cases. We refer to the result of a risk assessment finalised in 2012 by the European Commission's Scientific Committee for Consumer Safety (SCCS) on Phthalates in cosmetic products (SCCS 2007). Phthalates are a group of substances used in various applications. In cosmetics they were found as unintended contaminants. The opinion considered various phthalates. For one of them, the di-methyl phthalate DMP, the Committee found that "maternal toxicity" was the critical end-point from animal studies, with a NOAEL as indicated in the Table 4.1.

The Committee reasoned as reported below.

Margin of Safety (MoS) Calculation for Traces of DMP in cosmetics:

- A Margin of Safety, using the highest concentration (2982 mg/kg or 0.3 %) found, can be calculated, based on NOAEL and dermal absorption less than 5 % reported as a percentage in the literature.
- Only dermal exposure will be considered in the safety calculation. Maximum dermal absorption: <5 %
- Exposure to all cosmetic products: 17.79 g/day
- Default human body weight: 60 kg
- Systemic Exposure Dose (SED) $17.79 \times 1000 \times 0.3 \times 5\,\%/60 = 0.0445$ mg/kg
- No observed adverse effect level (NOAEL) = 600 mg/kg bw/day (maternal toxicity in rats)
- DMP Margin of Safety NOAEL/SED = 13,500

Conclusion DMP was found at concentration of 2982 mg/kg or 0.3 % in one perfume, in the other 15 perfumes the concentration was between 0.1 and 1.9 mg/kg. Using a worst case scenario, the Margin of Safety was found to be high. This suggests that unintentional exposure from perfume and other cosmetics at the levels found in this study would have no measurable risk for the consumer.

The conclusion was motivated by a margin of safety much above the minimum conventionally required level of 100, corresponding to a value of the safety/uncertainty factor of $10 \times 10 = 100$.

4.4.3.3 Method of Hazard Characterization of Non-threshold Effects of Chemicals

The Concept of Acceptable Dose Level For non-threshold effects, the aim of hazard characterisation is to set out a dose-level below which the extent of the effect is considered acceptable or to quantify the effect at a given level of exposure. In the case of non-threshold effects, the approach used aims at quantifying the risk at a given level of exposure, or to determine the maximum exposure corresponding to a pre-defined risk-level which is considered to be acceptable. For that aim, it is necessary to determine a dose- effect curve showing quantitatively the size of the effect as a function of the dose. We have already noted that, since in general only results from high-dose tests on animals are available, the part of the curve or mathematical relationship representing the dose-effect dependency at the much lower level of exposures, which are relevant for chronic human exposure, is based on extrapolation and is heavily dependent on the slope of the curve or mathematical model used for such extrapolation. Most often, one departure point is selected or calculated from the observed dose-response data obtained from the high-dose tests. From that point, a linear extrapolation is calculated, down to the dose levels in the human exposure range. In this case, interspecies differences may be covered by a factor which is incorporated into the extrapolation line. A specific value is incorporated into the extrapolation line for the interspecies factor related to toxico-kinetic aspects when it is possible to replace the external dose by the specific dose delivered to the critical organ or tissue, based on data from pharmacokinetic studies. Other aspects of variability, notably the human inter-individual variability, are in general not accounted for.

More sophisticated methods for hazard characterisation exist that are more biologically credible, but require data that in general is not available. Therefore, for non-threshold effects, notably genotoxic carcinogenicity, simplistic linear extrapolation is the most frequently used approach. It is a very rough approximation of the biological reality, which may be affected by large uncertainty.

Scientists have introduced an improvement to the methods described above as far as the establishment of a reference point is concerned. The improved method allows a better use of the animal test data by defining a benchmark dose as reference point.

The Concept of "Benchmark Dose" (BMD) As seen above the most used reference point for deriving a health- based guidance value is the NOAEL, the highest dose at which no statistically significant effect has been observed for the end points considered (relative to the reference, non-exposed group of test animals). Nevertheless, the NOAEL is only used for the so-called threshold effects. And the calculation of the NOAEL does not make use of the full set of dose-response data from the available animal studies. An alternative to the NOAEL, which can be used for both threshold and non-threshold data, is the "Benchmark Dose". The benchmark dose is a reference point (a dose value) which is calculated from the entire set of dose-response data available for the end-point considered. A dose-response curve is calculated by interpolation of the available data, from the point corresponding to the highest dose used, downward to the zero dose. An effect level is then chosen, normally the effect-level which is 5 % or 10 % above the incidence on the end-point considered that has been observed for the zero dose exposure. The benchmark dose is the lowest dose at which the response is expected with 95 % confidence level to be smaller than 5 % or 10 % above the incidence of the observed effect at zero exposure. The concept is illustrated in Fig. 4.2 (EFSA 2009). The observed data-points

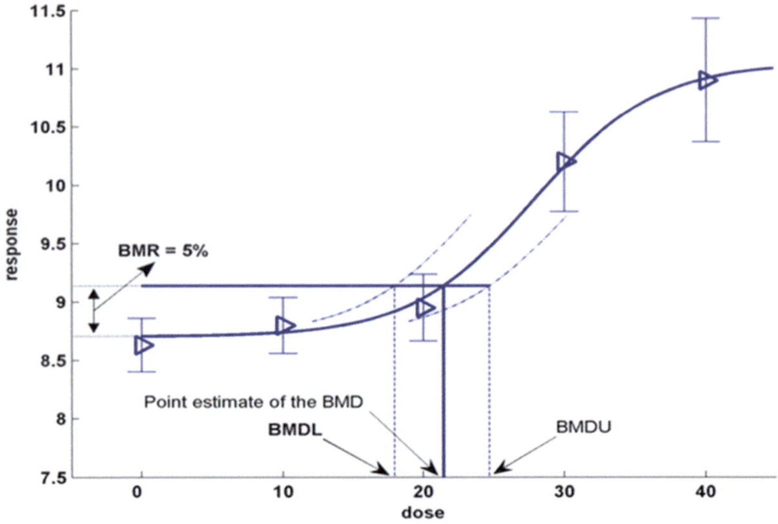

Fig. 4.2 Benchmark dose concept (Source: EFSA 2009)

are represented by the triangles. They are interpolated by the continuous curve. The upper and lower 95 % confidence bounds for the effect as a function of dose are represented by the dashed curves. BMD indicates the Benchmark Dose, while BMDL and BMDU are the lower and upper bounds, respectively, for the Benchmark Dose. These dose-levels are determined by the intersection between the horizontal line representing the 5 % effect-level above the effect observed in the reference group (zero dose) and the dose-effect curve.

Epidemiological Studies Few epidemiological studies are available and they have often their own limitations. Human data from epidemiological studies, when available, would provide more specific information overcoming the difficulties associated with extrapolation from high to low doses and from animals to humans. Nevertheless, these studies are rarely available. Moreover the few relevant ones are often affected by a lack of precision concerning the exposure of the different population groups, by the presence of confounding factors which influence the exposure-response relationship, and by the composition and differences in exposure and sensitivity of the exposed population.

In conclusion, hazard characterisation involves, in the absence of human data, extrapolation of dose-effect data from high dose animal tests to possible human adverse health effects at much lower doses. In the case of threshold effects, the uncertainty related to interspecies differences and the inter-individual variability are taken into account by applying conventional "safety" or "uncertainty" factors unless adequate data on toxico-kinetics and/or toxico-dynamics of the chemical considered allow the calculation of more specific factors. There is no complete agreement on the degree of conservativism of the conventional safety factors. In the case of non-threshold effects, the model or mathematical relationship used for extrapolating data from high dose animal tests introduces a high level of uncertainty. Interspecies difference is again taken into account via a conventional factor, while in general there is no systematic consideration of inter-individual variability in the current practice.

4.4.4 Step 3 of Risk Assessment: Exposure Assessment

Due to the pervasiveness of chemicals, consumers, workers and members of the general public are exposed, to different extent, to a great variety of substances, in different ways and situations. Consumers may be exposed to chemicals released by certain products and from the ingestion of contaminants, additives or other substances in food. Significant concentrations of chemicals may be present in work environments. The general population is exposed to a variety of pollutants, indoor or through the external environment, in the air, waters or through the soil. The exposure can take place through inhalation (for ex.: indoor and outdoor air pollution), ingestion (ex.: food) or dermal contact (ex.: cosmetics).

Once health hazards associated with a given chemical have been identified and characterised, exposure assessment aims to determine who is or can be exposed to it, in which way, at what level and for which duration. Exposure assessment is a key step in the assessment of health risks of chemicals. It is in many cases a complex exercise because of the variety of sources, the complexities of modelling the routes of exposure, the diversity of the population exposed and the problems posed by the possible concomitant presence and effects of chemicals other than the one under examination.

Exposure assessment is often a source of significant uncertainties. Risk managers should be aware of, and review with particular care the limitations and uncertainties associated with exposure assessment and their implications for risk characterisation.

Uncertainty and Variability There is an important conceptual difference between uncertainty and variability. Variability is an attribute of the object being studied, for example, there is variability in the body weight of individuals in a population in the sense that individuals have different body weights that the observer can however measure, at least in principle, and therefore describe quantitatively in a complete and accurate way.

Uncertainty is an attribute of our knowledge. It may be due to the state of development of scientific knowledge, the lack or the incompleteness and/or poor quality of data, or to the intrinsic randomness or unpredictability of the variable studied. Uncertainty can or cannot be resolved by research, data collection etc.; or can be resolved only in the long term.

Finally, there may be uncertainty on the variability. This is in particular the case of characteristics of a population that are measured on limited samples for practical or cost reasons. The inference from the sample to the entire population introduces a degree of uncertainty that depends on the sample size and representativeness. On the logical ground, the borderline between variability and uncertainty may however be difficult to define since much depends on the angle under which a phenomenon is considered, but for the practical needs of risk assessment it is not necessary to enter into a more theoretical discussion on this subject.

Assessing Exposure to a Chemical Involves Several Steps, Each Affected by Uncertainty First, one or more scenarios are defined in order to describe the possible sources of exposure and the situations and ways in which exposure can take place. WHO has defined exposure scenario as a combination of facts, assumptions, and inferences that define a discrete situation where exposure may occur. A scenario is the basis for building a quantitative model of exposure.

Scenario uncertainty refers to the possibility of overlooking, misrepresenting or incompletely describing possible sources, conditions and routes of exposure. It is particularly important when building an exposure scenario for the assessment of a chemical in a specific application or product to check whether other sources can contribute significantly to the exposure through the same or different routes (ingestion, inhalation, dermal contact, through environmental media, food, working environment etc.). If so, the effects of cumulative exposure should be considered or, if a comprehensive assessment is not practicable or is outside the mandate given to the assessors, decision makers should be made aware of the limited validity of the risk assessment. In such cases, at least an estimate of the possible contribution of other sources to the overall risk should be provided.

Second, a mathematical model is built in order to quantitatively link the sources of exposure to the target. The model may be more or less complex, depending on the configuration of the problem (number of sources, routes and possible chemical reactions in the transfer media, size and diversity of the target population etc.). Model uncertainty may relate to the form of the quantitative relationships between sources and targets as well as the parameters appearing in such relationships. Finally, data must be fed into the model, for instance the quantities of chemicals released over time, the duration of the exposure, the characterisation of the population exposed etc. Such data may also be affected by both variability and uncertainty.

Exposure assessment requires in particular the identification and characterisation of the population exposed. Age, gender, lifestyles, cultural habits, may influence very significantly the resulting exposure and the size and distribution of risk. The level of the exposure may indeed vary significantly between different populations and between groups and individuals within the same population. Dietary habits, for example, are different across countries, age and ethnic groups.

There is also a need to characterise the population according to its sensitivity or the expected response to the chemical considered, which again vary depending on body weight, age, sex, ethnicity etc. In practice, it is often very difficult or too time-consuming and costly to determine by direct data collection the variable characteristics of the populations and groups considered, and the range and distribution of exposure levels. Therefore, defaults values are used that may introduce approximation and uncertainty in the characterisation of risk.

Other sources of variability and uncertainty may be related to factors like the quantities of the chemical released during the use of the product or to the transportation, deposition, transformation, persistence, bioaccumulation or concentration of a chemical in the environment. In order to deal with the uncertainty and variability related to exposure assessment a stepwise, tiered approach is widely used.

Application of Tiered Approaches In a first step (Tier 0, screening analysis), very conservative hypothesis are made for the exposure and values that are at the higher extreme of the plausible ranges are assigned to all parameters, data and relationships. In such a way, a worst case is assessed. If the exposure and related risk resulting from such worst case assumptions is below the level considered safe or acceptable, there is no need for more sophisticated calculations.

If the results of Tier 0 do not allow excluding that the relevant thresholds are exceeded, or if for other reasons a more representative characterisation of the real exposure is needed, a Tier 1, qualitative analysis approach is applied. In such a case, the objective is to identify the main sources of uncertainty and variability, provide a qualitative description of their direction and their weight on the risk estimate, and conclude qualitatively on the likelihood of exceeding the level of risk that is considered safe or acceptable.

A further refinement is made with Tier 2 assessment that involves the application of sensitivity analysis to the most important parameters, relationships and data in order to identify and quantify the major sources of uncertainty and variability and calculate their influence on the resulting risks. The corresponding input values are modified within ranges considered to represent the possible uncertainty and/or variability, and the corresponding ranges of resulting risk are calculated, providing low, high or average predictions.

The most complex approach is represented by Tier 3 where probability distributions are assigned to the ranges of the most important input values and a resulting probability distribution is calculated for the resulting risk. The respective influence of uncertainty and variability on the resulting exposure and risk is sometimes calculated separately.

Exposure assessment may be simple in particular cases, like when the mode and quantity of the exposure to the substance considered are well determined (for instance when exposure takes place only through food, with well determined concentrations of the substance in food and a known distribution of dietary habits for the exposed population). However, it is most often complex and may be extremely difficult in cases like, for example, human exposure through the environment where it is necessary to model quantitatively the entire chain, from the source to the human targets, across various environmental compartments and geographical regions, calculating diffusion, transportation, chemical transformation, concentration, deposition, bioaccumulation etc.

4.4.5 Step 4 of Risk Assessment: Characterisation of Health Risk of Chemicals

Risk characterisation is the final stage of risk assessment. It combines exposure assessment with hazard characterisation. The objectives are to determine whether there is a health risk for exposed individuals, groups or populations, or part of them, what is the level of the risk and its distribution. In the presence of uncertainties, risk characterisation often also provides a qualitative or quantitative description of the probability associated with the level and distribution of risk.

Health risks of chemicals are expressed in different ways depending on the characterization of the underlying hazard. In the case of non-threshold effects like genotoxic carcinogenicity, risk is characterised in terms of increased incidence of the relevant pathology in the population exposed caused by the level of exposure considered. The risk may be modulated across different groups and a probability may be assigned to different levels of risk. In the case of threshold effects, risk characterisation informs whether the exposure of the population or particular groups is above or below the "safe" level. If the exposure is greater than the safe level, the distance to the safe level, quantified by calculating a safety margin, provides an indication of the level of the risk when the safety margin is too low.

Risk assessment may also be used to define the safe or the acceptable level of exposure to a dangerous chemical corresponding to the level below which it is very unlikely to observe appreciable risks (threshold chemicals), or the level corresponding to an acceptable risk (non-threshold chemicals).

The concept of acceptable risk, in particular when deciding which level of additional incidence of cancer in a population is acceptable, poses important principle problems. Figures like, for example, one in a million or one in 10,000 are sometimes used. However, there is no EU political decision endorsing the use of any such figure. Quantification of the risk is useful, but decisions on what is "acceptable" should be left to political decision makers, with a clear, transparent distinction between technical assessment and policy decisions. More will be discussed on this issue in Chap. 5 on Risk Management.

4.4.5.1 An Example of Risk Characterisation: DEHP in Medical Devices

A simple example of risk characterisation is provided by an opinion on the safety of medical devices containing DEHP-plasticized PVC or other plasticizers, adopted by the European Commission's Scientific Committee on Emerging and Newly Identified Health Risks (SCENIHR 2014).

PVC is used in parts of certain medical devices intended for procedures such as blood transfusion, hemodialysis, parenteral nutrition or endotracheal tubing. In order to ensure the necessary flexibility of the relevant parts of a medical device, various plasticizer substances are used, among which the most frequently used is DEHP [di-(2-ethylhexyl) phthalate]. Since DEHP may migrate from the PVC into the human body, patients are exposed as a result of the medical practice.

SCENIHR has identified and characterized the relevant hazards of DEHP by a review of the literature and documents reporting test results. It has concluded that the target organs of chronic toxicity are the liver, the kidney and the testis. Among the NOAELs related to the various adverse health effects identified for DEHP, SCENIHR has selected the lowest one, which is that related to testicular and developmental toxicity. The value so selected for the NOAEL, 4.8 mg/kg bw/day, was reduced by a factor 100 (10 × 10) in order to determine a Tolerable Daily Intake (TDI), that is a level of exposure below which the likelihood of adverse effects in humans is considered negligible. This approach is justified by the fact that the identified hazards of DEHP consist of threshold effects. The TDI obtained is 48 μg/kg bw/day. This TDI is a useful reference value for characterizing the risk of chronic exposure to DEHP. From the available data, the SCENIHR has calculated the exposure of different categories of patients, resulting from medical procedures using DEHP-containing devices.

The SCENIHR has then reached the conclusion that adult patients undergoing chronic hemodialysis treatment had the highest exposure because long-term hemodialysis as a continuously repeated procedure was liable to result in the highest cumulative dose of DEHP.

Since the TDI is a reference value for chronic toxicity it was appropriate to use it for the hemodialysis scenario. The SCENIHR observed that the median exposure levels exceed the TDI by 2 to 12-fold with peak values >40-fold higher than the TDI."

The SCENIHR has also considered another adverse health effect of DEHP, kidney toxicity, concluding that the exposure values had a small Margin of Safety in relation to the NOAEL in rodents for that end-point. The overall conclusion was that patients subjected to hemodialysis procedure were at risk of DEHP-induced effects.

4.5 Issues in Health Risk Assessment of Chemicals

We address now some of the critical issues on the health risk assessment methodology for chemicals that require attention and a critical understanding by risk managers and interested stakeholders in order to interpret properly risk assessment findings, and to interact productively with risk assessors.

4.5.1 The Importance of Properly Defining the Scope of Health Risk Assessment

The scope of a risk assessment should be clearly defined by managers. Should assessors consider exposure solely from the product considered or should they take into account, as far as possible, other sources of exposure? Should they take into account the possible concurrent effects of other chemicals? Should they consider only the direct consumer exposure or also exposure of workers at the production stage and of the general population through the environment at the end-of-life stage? The definition of the scope of the assessment has a substantive impact on the meaning and validity of the results and should be carefully considered and transparently communicated.

Often, for practical reasons or in view of the requirements of the relevant legal framework, a predefined and limited scope is assigned to the assessment. For instance, assessors may be requested to focus the assessment on the risks associated with the use of a household product by consumers, with no regard to the possible health impact for workers and for the public, respectively at the production and disposal stages. In such cases, risk assessment provides only a partial view of the risks in presence. Equally, risk assessment of a chemical in a specific product provides only a partial view of the overall risk if other sources of exposure to the same chemical are not taken into account (aggregated exposure). In certain cases the concomitant and similar effects of other chemicals (cumulative effects) may also need to be taken into account in order to provide a realistic assessment of the risk.

Therefore, risk managers, when requesting health risk assessment advice, should define clearly the objectives and scope of the assessment and be aware of their implications. Risk assessors should communicate in a transparent manner, and bring to the attention of decision makers and the other users of the risk assessment, any limitation of the validity of their conclusions, which may depend on the way in which the scope of their assessment was defined. Some additional considerations on this issue are presented further on in this Chapter.

4.5.2 The Conventional Assumptions Behind Risk Quantification

The result of a risk assessment is often expressed in terms of tolerability, acceptability of the risk studied, or of "safety" related to technical parameters like "margins of safety". We wish to raise the reader's awareness to the need for risk managers and decision makers to carefully review the criteria used for quantifying the level of risk, something on which we will insist again in Chap. 5. Conventional safety factors and other criteria are used in risk assessment for grading risks as "acceptable", "tolerable" or "negligible":

Definition of Tolerable Daily Intake-TDI A TDI may be defined as the maximum daily intake of a substance that is not expected to cause any appreciable health risk over a lifetime. TDIs are used for assessing the risks and determining the regulatory requirements for substances that are not intentionally added to food. The intake may be caused by exposure through the air, water, food or any other route. TDIs are determined applying uncertainty/safety factors to data obtained from toxicity studies on test animals. They are mostly used for food risk assessment and management.

Definition of Acceptable Daily Intake-ADI An ADI may be defined as the maximum amount of a substance in food or drinking water that can be ingested daily over a lifetime without an appreciable health risk. There is no substantial difference between the ways in which TDIs and ADIs are calculated. The difference between tolerable and acceptable refers to the non-intended or intended presence of the substance. Nevertheless, the distinction is not always respected in practice.

TDIs and ADIs are expressed usually in milligrams per kilograms of body weight per day.

4.5.3 The New Perspectives for the Future of Risk Assessment of Chemicals

It is important to also note that current methods are not final and the scientific basis of risk assessment is rapidly evolving notably due to progress in molecular biology and computing. Health risk assessment is today performed in most cases by conventional methods that have not changed significantly since many decades. Nevertheless, recent scientific developments, notably those referred to as "omics", which have not yet been incorporated into systematic practice, open new promising perspectives in the longer term, but pose also new challenges for the future of health risk assessment.

What Are "omics"? Three techniques deriving from recent developments of biology and computer technology are of particular interest for health risk assessment: genomics, proteomics and metabolomics. The three are characterised by the suffix "omics" that indicate according to the English Oxford dictionary "a totality of some sort", underlining that these techniques rely on the capacity to perform collection and analysis at once of data on a very large scale

Genomics is based on the ability to analyse the structure and function of genomes by use of high-throughput sequencing that allows to read large numbers of DNA sequences at once (in the order up to millions), as well as the profiling of gene expression (transcriptomics), and of epigenetic modifications on the genetic material (*epigenomics*).

Proteomics techniques allow the reading and analysis of expression by the genes of proteins of the cell and tissues.

Metabolomics allows the identification and quantification of metabolites (products formed in the organisms in the process of degradation or elimination of the original chemical) in a biological system, the cells, tissues, organs or an entire organism.

The interest of "omics" for the assessment of chemicals is that these techniques allow the detection of the effects of substances on certain biological parameters at cell or molecular level with a high degree of sensitivity, short exposure times in the case of animal tests or in in-vitro tests, at dose-levels more representative of human exposure compared to the traditional animal tests. In fact, compared to the classical toxicological tests, "omics" provide closer insights into mechanisms of action which may be related to possible hazard characteristics of a chemical.

In short, the current interest of "omics" techniques for risk assessment relates to their potential to facilitate and improve hazard identification by contributing to the understanding of modes of action of chemicals on biological systems. The increased sensitivity allowed by "omics" techniques in detecting biological effects is an important advantage but poses difficult problems of interpretation and assessment of their health significance and quantification of health effects.

4.5.4 Aggregate Exposure

As we have already noted, depending on the specific legal framework applicable, risk assessors are often requested to assess the risk posed by the use of a given chemical in a specific product or application. When the assessment mandate is formulated in a narrow way, the result of the assessment may be restricted to the risk that the chemical in question poses when the exposure is exclusively due to the application considered. Nevertheless, consumers, workers or the general public may also be exposed to the same chemical from sources other than those considered in a narrow approach. For example, a certain chemical substance may be used in food-contact material, but also in toys or cosmetics, like in the case of certain phthalates. Those different applications are regulated under different legislative instruments. It may happen that the use of the substance in food-contact material up to a certain concentration is found to be "safe" when considered in isolation. Therefore, the use of the substance below such safe concentration may be allowed under the specific legislation on food-contact materials without consideration of the possible parallel exposure from toys and cosmetics.

An Integrated Approach to Aggregated Exposure In the case of multiple sources of exposure to a given chemical, an integrated approach to a chemical may be necessary in cases where the aggregated level of exposure from the various sources is significant. Such integrated approach may require collaboration or coordination between different assessment bodies. It should be in the first instance the responsibility of risk managers, when consulting a risk assessment body, to request a comprehensive review of the possible sources in order to determine the need for an integrated assessment. Sometimes an integrated assessment may be difficult to perform because of the lack of data on particular sources of exposure or on the route or extent of the exposure from sources other than the regulated application considered. In such a case, at least a qualitative analysis should be made and the risk assessment report should draw the attention of risk managers to the limited validity of the results presented.

4.5.5 Cumulative Risk Assessment of Mixtures

The assessment methods that we have introduced in this Chapter consider in general the potential adverse health impact of a specific chemical in isolation. Nevertheless, consumers, workers and the general population are normally exposed to a variety of chemicals and to other non-chemical stress factors. It is therefore important for the effectiveness of protection measures to consider whether the various substances and factors in presence may give rise to cumulative or synergic adverse health effects. Cumulative risk assessment is the assessment of the combined risk deriving from the aggregated exposure to multiple chemicals and stressors. This is clearly a very complex issue. In fact, as yet there is no generally recognised approach to cumulative risk assessment. We focus here on the cumulative assessment of chemical mixtures.

Two alternative hypotheses are considered by scientists:

(a) *Independent action*: The effects of different chemicals are considered to be additive only if the chemicals considered produce a common adverse effect on the same target (organ, system) with a similar mode of action. The mode of action (MoA) is the biological mechanism through which the chemical molecules act in the organ and induce the effect. Under this hypothesis, the chemicals with a dissimilar mode of action are assumed to act independently and therefore the approach is labelled as "independent action" (IA). In the case of chemicals with a similar mode of action, even if the level of exposure to each chemical taken individually is lower the "safe" level, there is a risk when the cumulated exposure exceeds such level.
(b) *Dose addition*: The effects of chemicals that produce the same effect on the same target should be treated as additional (Dose Addition, DA) even if they have a dissimilar mode of action.

A fundamental problem for cumulative risk assessment derives from the practical difficulty of testing the toxicological properties of the enormous variety of mixtures to which we may be exposed. Moreover, data on modes of action often are not available and in any case it is difficult to determine whether different modes of action are really mutually independent.

The European Commission Communication on "The combination effects of chemicals, Chemical mixtures" The debate on cumulative risk assessment is not yet closed. The European Commission has addressed in 2012 this difficult and important issue in its Communication on "The combination effects of chemicals, Chemical mixtures" (European Commission 2012). The Communication was the response to an invitation of the Council

> … to assess how and whether relevant existing Community legislation adequately addresses risks from exposure to multiple chemicals from different sources and pathways, and on this basis to consider appropriate modifications, guidelines and assessment methods, and report back to the Council by early 2012 at the latest.

In its Communication the Commission relied on a Joint Opinion of its three Scientific Committees, SCHER, SCENHIR and SCCS (SCHER, SCENIHR, SCCS 2012). On the regulatory situation, the Commission concluded that:

> …where chemical mixtures are assessed and regulated under EU legislation this is predominantly in relation to chemical products of known composition. Some complex mixtures discharged/emitted to the environment from a single source (factory, facility, etc.), are also subject to controls. There are a few examples of assessments and controls being carried out in relation to several substances originating from different sources and through different pathways but these are limited in their scope. Currently, within the framework of EU legislation, there is no mechanism for a systematic, comprehensive and integrated assessment of mixture effects taking into account different routes of exposure and different product types.

The Opinion of the Commission Scientific Committees on Chemical Mixtures On the assessment of mixtures, the main conclusions emerging from the Joint opinion of the three Scientific Committees were:

1. Under certain conditions, chemicals will act jointly in a way that the overall level of toxicity is affected.
2. Chemicals with common modes of action will act jointly to produce combination effects that are larger than the effects of each mixture component applied singly. These effects can be described by dose/concentration addition.
3. For chemicals with different modes of action (independently acting), no robust evidence is available that exposure to a mixture of such substances is of health or environmental concern if the individual chemicals are present at or below their zero effect levels.
4. Interactions (including antagonism, potentiation, and synergies) usually occur at medium or high dose levels (relative to the lowest effect levels). At low exposure levels, they are either unlikely to occur or are toxicologically insignificant.
5. In view of the almost infinite number of possible combinations of chemicals to which humans and environmental species are exposed, some form of initial filter to allow a focus on mixtures of potential concern is necessary. Several criteria for such screening are offered.
6. With regard to the assessment of chemical mixtures, a major knowledge gap at the present time is the lack of exposure information and the rather limited number of chemicals for which there is sufficient information on their mode of action. Currently, there is neither an agreed inventory of mode of actions, nor a defined set of criteria how to characterise or predict a mode of action for data-poor chemicals.
7. If no mode of action information is available, the dose/concentration addition method should be preferred over the independent action approach. Prediction of possible interaction requires expert judgement and hence needs to be considered on a case-by-case basis.

In short, the three Committees pronounced in favour of the independent action approach for chemicals acting with dissimilar mode of action on the same end point, as far as human health was concerned, stating however that in the absence of specific information on mode of action the dose addition approach had to be preferred as a measure of caution.

In another part of their opinion, the three Committees were more specific on ecological aspects stating that for ecological effects the exposure to mixtures of dissimilarly acting substances may raise a concern even if all substances are below the individual **P**redicted **N**o **E**ffects **C**oncentrations (PNECs). The Committee concluded that an effort should be made in order to develop a better knowledge and improve the methodologies in view of comprehensive approaches for the ecological risk assessment of chemicals under realistic conditions.

The Opinion of the EFSA Panel on Plant Protection Products However, 2 years after the adoption of the Joint Opinion of the three committees, in an opinion on the relevance of dissimilar modes of action for cumulative risk assessment of pesticides residues in food (EFSA 2013), the EFSA Panel on Plant Protection Products (PPR) reached a significantly different conclusion as far as mixtures of pesticide residues were concerned stating that based on the available knowledge and account taken of the relevant uncertainties, there were great conceptual and practical difficulties in making a distinction between similar and dissimilar Mode of Action and in deciding which concept (dose addition or independent action) should be used for risk assessment practice. It recommended to use cumulative risk assessment methods derived from dose addition also for the assessment of mixtures of pesticides with dissimilar modes of action, provided they produce a common adverse outcome.

Divergent Opinions Between Scientists on the Way to Assess Chemical Mixtures The divergence between the two opinions is a good example of what we abundantly highlight in this book: risk assessment is not pure science; its results depend on many assumptions, decisions on the weight of various types and lines of evidence, subjective judgements etc. Risk managers are often called to resolve uncertainties and try to reconcile divergent opinions. Moreover, good practice requires that important divergences are not left unnoticed by risk managers. More collaboration and joint assessments between risk assessment bodies might help reconciling divergent opinions or at least clarifying the reasons for the divergent conclusions.

Finally, consideration of synergisms between chemicals and of cumulative/synergic effects of stressors of different nature are still matters for research and are not expected to enter the applicative stage in the near future. Risk managers need therefore to be aware that synergic, cumulative or multi-stressors effects, which may be significant under certain specific circumstances, are often disregarded because of lack of data and/or the insufficient development of assessment methodologies.

4.5.6 *The Threshold of Toxicological Concern*

Determination of the toxicological profile of a chemical requires in most cases costly and time-consuming studies involving tests on animals. In the case of very low level exposure to chemicals for which specific data are not available, a method has been developed that allows to determine a "safe" level of exposure, which is extrapolated from experimental toxicological "points of departure" like NOAELs of a large set of chemicals. The chemicals are grouped in classes according to the aspects of their structure that are relevant in relation to their potential toxicity. The statistical distribution or the NOAELs for each class of chemicals is calculated and the NOAEL value below which fall no more than the 5 % of the observed NOAELs is determined. Then, a threshold of toxicological concern (TTC) is calculated dividing by a safety factor (normally 100) the value so determined.

The assumption is that any exposure to a chemical with a certain structure, lower than the threshold of toxicological concern for the corresponding class may be considered "safe". In other terms, such lower exposure is considered unlikely to pose a significant health risk.

The TTC approach is used in a limited number of sectors when there is no requirement for specific toxicological data. It is a powerful method for determining priorities for tests, focusing on those chemicals for which the exposure is higher than the corresponding TTC.

4.5.7 *Assessment of Carcinogenic Substances*

As already noted, carcinogenic substances are divided, for the needs of risk assessment, into two classes depending on the mechanism of carcinogenesis: genotoxic and non-genotoxic. We have seen that non-genotoxic substances are assumed to show a threshold for their effects and therefore are treated as other "threshold chemicals" by identifying a point of departure and applying safety factors. In the case of genotoxic and carcinogenic chemicals, it is assumed that no such threshold exists. An extrapolation from the high dose animal data to the much lower dose levels representative of human exposure conditions allows to establish a level of exposure corresponding to a low risk considered acceptable (increased cancer risk in a lifetime or increased number of additional cancer in a population exposed over life time), or to calculate the additional number of cancer at a given level of exposure.

The Margin of Exposure Approach to the Assessment of Genotoxic Carcinogenic Substances Other approaches have been developed like in particular the margin of exposure calculation. Under this proposed approach, a point of departure is established from the available animal data. Such point of departure may be calculated for instance with the benchmark approach. Then an acceptable exposure value is calculated by applying a "margin of exposure" factor normally of 10,000 corresponding to the usual $10 \times 10 = 100$ safety/uncertainty factors multiplied by a further factor

100 to take into account the uncertainties related to the carcinogenicity mechanisms. The conventional nature of the methods is self-evident.

4.5.8 The Crucial Importance of Uncertainty and of Its Expression in Risk Assessment

As discussed in Chap. 2 on the definition of risks, uncertainty is intrinsic in risk. Risk is basically about uncertainty. Uncertainty plays a great role in the risk assessment of chemicals. It should be treated systematically and expressed in a transparent way. It is therefore useful to examine more closely the different types of uncertainties in health risk assessment of chemicals, which are relevant for risk assessment in general.

Uncertainty in Health Risk Assessment of Chemicals Uncertainty refers to imperfect knowledge of the present or future state of a system or of the outcome of an event. The system may be a biological system, like in the case of human health an organism, an ecosystem or a technological or human-built system (for instance, a financial system).

The assessment of health risks of chemicals requires a great deal of substance and source-specific data as well as data on exposure and on the effects of the chemical studied on the human organisms, which may be significantly affected by uncertainty. We have already mentioned that there are different types of uncertainties in relation to the data and knowledge required for characterising health risk of chemicals. In summary:

Epistemic Uncertainty Uncertainty due to lack for practical reasons of adequate and specific data or knowledge. Examples are the lack of specific data on body weight distribution or dietary habits in a population, the concentration of an air pollutant across a given region or the rate of bioaccumulation of a substance in certain species. All this information could in principle be obtained by specific measurements, which however are often not feasible in practice or too time consuming or costly. In such cases, defaults or data from limited samples are used, which introduce a certain degree of uncertainty. Since this type of uncertainty could in principle be reduced or eliminated by further data collection, tests, studies etc., it is important to inform decision makers of the nature of such uncertainty, as well as on the feasibility, conditions and time scale for possible knowledge improvements.

Uncertainty may also derive from the methodologies used to produce certain data, in particular the use of animal models for toxicity testing. Further limitations may be introduced by restrictions on animal testing itself. This type of uncertainty could only be removed through scientific breakthroughs in toxicology. New developments in molecular biology and increased computational power are opening promising perspectives for improvement in that respect.

"Completeness" Uncertainty Completeness uncertainty derives from the possible ignorance of facts, phenomena or mechanisms that may have an impact on health, but have not (yet) been identified and studied. These are "unknown unknowns" that by definition cannot be taken into account, but may suggest precaution notably in relation to risks of new technologies, when the scientific knowledge is still limited and evolving and the possible negative consequences of "unknowns" may be massive and/or irreversible.

Aleatory Uncertainty Other uncertainties are caused by the randomness or unpredictability of the underlying phenomena or processes. This type of uncertainty may apply to physical and biological systems but also to human behaviour and the evolution of human systems. This type of uncertainties cannot be eliminated, but can possibly be treated probabilistically.

There are several ways of consolidating and expressing the various types of uncertainties and variability that affect risk assessment:

A Qualitative Approach to Uncertainty and Variability A simple example of a qualitative (Tier 1) approach based on consideration of the direction of uncertainties is provided in tabular form in Table 4.2 (modified from a tabular form proposed by the European Chemical Agency (ECHA 2012-1)):

Table 4.2 Example of table for the qualitative assessment of uncertainties

	What is affected	Cause of uncertainty	Variability or uncertainty	Direction and magnitude
Hazard assessment	Model	Cause 1	VAR	–
	Input parameters	Cause 2	UNC	+++
		Cause n	UNC	++/––
	Overall effect on hazard estimate			
	e.g.: Likely overestimation of hazard due to Cause 2. Possible to reduce uncertainty by…			
Exposure assessment	Scenario	Cause 1	UNC	++
	Model	Cause 2	VAR	+
		Cause 3	UNC	+/–
	Input parameters	Cause 4	UNC	–
		Cause m	VAR	––
	Overall effect on exposure estimate			
	e.g.: Likely overestimation due to Cause 1 and Cause 2. Uncertainty may be reduced by …			
Risk characterization	Overall effect on risk estimate			
	e.g.: the risk estimate is likely to be overestimated mainly based on assumptions in exposure assessment. Overall uncertainty may be reduced by…			

VAR variability, *UNC* uncertainty
+, ++, +++ = low, moderate and high overestimates; –, ––, ––– = low, moderate and high underestimates

With this methods, risk assessors provide qualitative information on the influence of the various sources of uncertainty or variability on the assessment of hazard, exposure and the resulting risk. Such a structured form of presentation should however not induce a false sense of precision since it is based on subjective and qualitative appraisal of the underlining uncertainties.

A Probabilistic Representation of Uncertainties and Variability We have seen that a more sophisticated quantitative approach may be applied when an overall distribution of probabilities or the resulting risk may be produced by compounding randomly-assigned values from the probability distributions of the underlying causes of uncertainties and variability. The overall distribution may represent separately uncertainty and variability. This type of representation may be justified when it is possible to estimate quantitatively the underlying distributions of probabilities with a sufficient degree of confidence. Otherwise, it may again induce a false sense of precision. Moreover, the meaning, the basis and the way of calculation as well as the limitations of such quantitative distributions are not easily understood by those decision makers who do not have a scientific background. Probability distributions presented in risk assessment reports should therefore be accompanied by an adequate narrative enabling the reader to properly interpret the quantitative information. Unfortunately, there is no general consensus on the best way to communicate uncertainties. Use of qualitative terms is subject to ambiguity, while a quantitative expression may give a false sense of precision and does not necessarily support a consistent appraisal. We will come back to this issue in Chap. 8, describing an approach that combines qualitative and quantitative information.

4.5.9 Review of Information and Data and the Weight of Evidence Approach

Risk assessment is basically a process of evaluation and elaboration of information and data aimed at identifying and characterising risks. The starting point of the process is therefore the identification, evaluation and selection of the relevant information and data. Regulatory risk assessment often starts with the evaluation of information and data submitted by the applicant for an authorisation or by the proponent of a project. However, for the credibility of the process, the applicant's dossier should be evaluated taking into account any other relevant data that the assessors may collect or be aware of. In fact, a comprehensive review of the available knowledge, information and data should be considered an important precondition for high quality and credible advice on risks. Although this may be seen as a trivial, common sense statement, its practical application is far from easy in practice. How data are sourced, selected and evaluated may become a cause for controversy on the validity and acceptance of the risk assessment conclusions.

Steps of Data and Information Sourcing and Assessment The first important step is the appropriate formulation and framing of the risk issue and of the questions to be submitted to risk assessors. Assessors need to identify search terms and key words for a literature search. A basic condition for the quality and credibility of an assessment report is that it documents the way in which the data search has been organised: which subjects, terms and key words have been addressed, which tools have been applied in order to get access to data and information, what types of sources have been considered.

The second step is the selection of sources and types of documents: published, peer reviewed reports, other publicly available reports, informal papers, personal communications, stakeholder submissions, confidential data submitted by industry etc. The final step is the decision on what information and data are taken into account and how they are "weighed", that is: how far the various data available are allowed to contribute to the conclusions. Here come into play the weighing criteria, for instance the criteria applied for deciding on the validity of the test methods used to produce the data.

Weighing Data and Information from Different Lines of Research and of Different Quality Risk assessment, health risk assessment of chemicals but also many other types of risk assessment, are rarely based on complete, specific, relevant knowledge and data of unchallengeable quality. In many cases, knowledge and data are characterised, to a certain extent, by incompleteness, uncertainty, inconsistency or limited quality. Risk assessors need to put together and integrate "evidence" from different lines of research, different sources, different methods etc. They need to integrate data in a weighed manner, considering how each available data-set contributes to reinforce or, on the contrary, weaken a given hypothesis. Therefore, in many risk assessment reports, the conclusions are presented as the result of a "weight of evidence" approach. This is an aspect of risk assessment where the professional judgement of assessors plays a very important part.

Weight of Evidence Guidelines EU risk assessment bodies and committees have defined guidelines (ECHA 2010, 2012-2; SCENIHR 2012) aimed at ensuring consistency and transparency in the application and reporting of the weight of evidence approach. Although such guidelines reflect practices which are widely recognised by the risk assessment community, it should be observed that "weight of evidence" is a key aspect of risk assessment policy that would deserve a more formal definition in collaboration between managers and assessors.

Some common aspects of the weight of evidence guidelines currently recommended by the EU risk assessment bodies can be synthesized as follows:

- A weight of evidence approach is used when, in the absence of specific, high quality information and data, it is possible to reach conclusions pooling together information and data of a lesser quality;
- The available information is evaluated and weighed for its reliability, relevance, adequacy and quantity:

- Reliability expresses the quality of the study or research producing the information considered. It includes the validity of the methodology used, the good design of the study/research, and how well the study/research is conducted and reported. A score proposed by Klimisch is often used (Klimisch et al. 1997):
  - 1 = Reliable without restrictions
  - 2 = Reliable with restrictions (some deficiencies in the design and/or conduct and/or reporting)
  - 3 = Not reliable (serious concerns about the design or conduct of the study/research)
  - 4 = Not assignable (not possible to make an evaluation because information provided on method/design/conduct is insufficient)
- Relevance expresses the extent to which the information produced by the study/research considered is appropriate for making a judgement on the issue considered (for instance the stress agent, the hazard etc.). The relevance may be direct (specifically related to issue), indirect (not specific but transposable) or insufficient;
- Adequacy/Utility expresses how far the information/data provided can contribute to the assessment of the aspect considered (for instance to characterise the hazard or to quantify the exposure). That depends on the content of the study;
- Quantity expresses the amount of information provided.

In the overall evaluation, the different types of evidence available are weighed and integrated. In the case of chemical health risks, lines of evidence for the toxicological aspects are often categorised as follows:

- Animal studies
- In vitro studies
- Epidemiological studies
- Mechanistic/mode of action studies
- Mathematical modelling
- Studies of toxicokinetics

Different types (lines) of evidence may be considered for other aspects of the assessment. For instance, as far as exposure is concerned, there may be information from measurements on a sample-basis or from mathematical modelling.

Need for Transparency on the Weighing of Evidence Although the indicative criteria that may be found in guidelines help promoting consistency, the final weighing and integration of evidence remains largely a matter of judgment. Transparency on the approach and on the specific criteria applied is therefore essential for the credibility of a risk assessment report. This is why the existing weight of evidence guidelines or guidance documents of certain EU risk assessment bodies also include specific provisions on how to report on weight of evidence.

Finally, it should be noted that the application of the weight of evidence approach is sometimes criticised as a grey aspect in regulatory risk assessment. The exclusion of, or the low credit assigned to studies not applying validated methods or not complying with internationally recognised good laboratory practices is sometimes seen as a way to discriminate against inconvenient evidence challenging the validity of mainstream conclusions. Obviously, the relatively large room for subjective judgement open by the weight of evidence approach attracts the suspicion of bias in case of conclusions which are unwelcome by part of the stakeholders.

The Tension Between Rigour and Precaution in Weighing Evidence The absence of a more formal, consistent and recognised approach to the weighing of evidence does not help in that respect. It is likely to be difficult, even by some specific research, to draw conclusions on whether the current practice of weight of evidence works in favour or against precaution and protection.

It can be said that there is a tension between the scientific and the "political" approach to weight of evidence. The scientific method tends to minimise the risk of the so-called Type I errors, that is: accepting as true hypotheses that in the end come out to be false (for instance concluding that a substance if carcinogenic when it is not). The political approach is instead keener to avoid Type II errors, that is: rejecting hypotheses which in the end are found to be true. The reason is clear: science is afraid of endorsing wrong theories whereas political management is afraid of bearing the responsibility of health and environmental damage or deaths. Unfortunately, what is gained in scientific rigour is lost in precaution. The criteria used for weighing evidence play an important part in this zero-sum game.

4.5.10 Causality in Risk Assessment

A basic task in risk assessment is to verify hypotheses about causality links between possible stress factors and observed adverse effects. We have seen that in the case of animal testing, a judgment is made on the toxicity of a chemical by comparing the incidence of a certain adverse effect in the exposed group to that in the reference, non-exposed group of test animals. The judgment is based on probabilistic considerations. A causality link between exposure and effect is accepted if the probability of the observed increase in the incidence is less than, say, 5 % or 1 % in the case of a difference only due to the intrinsic variability of the adverse effect in the type of animals used.

More in general, discriminating between simple association and causality is not always an easy task. Association may be grossly misleading, bringing us to interpret A as the cause of B just because they go often together. This is an issue of particular relevance in epidemiological research.

The Bradford Hill Criteria for Causality Some classical, authoritative criteria have been suggested in 1965 by Sir Austin Bradford Hill, a distinguished professor of medical statistics at the University of London. The Bradford Hill criteria (Bradford Hill 1965) are still considered very helpful for the critical examination of association and causality. According to those criteria, an association observed between event A (the possible cause) and the event B (the effect) should be examined under nine aspects:

- *Strength*. A causal link may be considered more likely if there is a strong association between A and B, that is if B follows A in a great number of cases.
- *Consistency*. The causal hypothesis is reinforced when the association is observed in a variety of places, times, circumstances.
- *Specificity*. If B follows A in a very specific way, meaning that event B is nearly exclusively associated with A, the causality link may be considered more likely. Nevertheless, the opposite is not true. The absence of specificity does not automatically justify the exclusion of causation. Sir Bradford Hill observes that multi-causation is more likely, in general, than single causation.
- *Temporality*. In order to conclude that A is the cause of B, B must appear after or in concomitance with A. This criterion helps examining whether A is the cause of B or B is the cause of A. Bradford Hill makes the example of a diet associated with a certain pathology: does the pathology follow in time the diet, or is the diet embraced as a consequence of the pathology?
- *Biological gradient*. If there is a quantitative correlation between A and B, a dose-response curve showing a higher incidence of B when A increases, the hypothesis of a causality link is reinforced.
- *Plausibility*. Causation may be considered more likely if there is a plausible biological explanation for the link between A and B (more in general a plausible explanation based on the established relevant scientific knowledge). Obviously, since science is continuously evolving, this cannot be an absolute criterion.
- *Coherence*. The cause-effect interpretation should not be seriously incompatible with well-established facts and scientific knowledge. Although plausibility is not an absolute criterion, conflict with basic scientific knowledge undermines the causality hypothesis. In other terms, it may be possible to accept causality even if we cannot explain, based on current knowledge, how the link works, but it become more difficult to accept it if a causal link contradicts well established facts and knowledge.
- *Experiment*. When it is feasible, an experiment showing that removing A removes B would more clearly confirm the causal link.
- *Analogy*. Relevant precedents may reinforce the causal hypothesis. Sir Bradford Hill mentions the experience of thalidomide effects as inviting to caution in judging even weaker associations between drugs and adverse effects in pregnancy.

None of these criteria has an absolute value and any association should be examined in a comprehensive and balanced way, in light of all of them. Sir Bradford Hill refers to the need for caution in considering also what is at stake. Our discussion of Type I and Type II errors is relevant in that respect. Accepting or rejecting causation may have huge consequences for health and the environment as well as important economic impacts. This is not just a matter for professional judgement. Although often the issues involved are complex, transparency is key and risk assessors should make efforts to allow risk managers and stakeholders to understand the reasoning behind the conclusions proposed.

4.5.11 The Assessment of Endocrine Disrupting Substances

The assessment and regulation of endocrine disrupting substances is one of the current most difficult and controversial issues of chemical safety. It shows the importance (and difficulties) of the distinction between hazard and risk, biological effects and adverse health effects, threshold and non-threshold effects. EU regulation on such substances has been shaped by precautionary considerations but its application and effectiveness depend on clear definitions and appropriate tests covering all the relevant end-points. Scientists are divided on the way to define and assess endocrine disrupting substances. The potential practical, industrial and cost-impacts of different definitions and management approaches are huge. Interests have mobilized their experts and scientists have been drawn into non-scientific arguments with a worrying confusion of roles. The European Commission has been brought to the front line in trying to arbitrate and resolve scientific issues on which scientists themselves do not seem able to reach consensus.

The Endocrine System The endocrine system is one of the vital systems of the organism, in humans and animals. It is constituted by a group of glands secreting hormones directly into the circulatory system. Hormones are not a homogeneous class of chemicals: there are a wide variety of chemicals with different structure that are categorized as hormones because of their functions in the organism. Therefore, "hormone" is a functional, not a chemical category.

The hormones are carried to tissues and organs across the organism. They serve as messengers to cellular receptors and regulate important processes in the development and the functioning of the organism. This is why the system is also called the endocrine-signaling system. Hormonal signaling involves several steps starting with the biosynthesis of a particular hormone in a gland, the storage and subsequent secretion of the hormone, the transport of the hormone in the blood stream to the target cell, the recognition of the hormone by a receptor protein in the cell membrane or inside the cell and finally the cellular response. The signal may be amplified and there are feedback mechanisms allowing the regulation of the production of hormones as a function of the reaction in the target cells. This is a form of "homeostasis" aimed at modulating the production of hormones via a negative feedback loop. Hormones intervene also in regulating the production of other hormones. Finally the hormones break down and are eliminated.

The receptors are protein that may be located on the membrane or inside the cells. The most important components of the system include in particular the pineal ad pituitary glands, the pancreas, ovaries, testes, thyroid, parathyroid, hypothalamus, gastrointestinal tract and adrenal glands. Several other tissues are also involved in producing hormones. It is considered that the endocrine system is organized around three main axis: the hypothalamus-pituitary-gonad axis, the hypothalamic-pituitary-adrenal axis and the hypothalamic-pituitary-thyroid axis. In reality, both the composition and the functioning of the endocrine system are extremely complex. Endocrinology is a self-standing, well established branch of medical sciences. Nevertheless, relevant knowledge is progressing fast. There is discussion on how to define the endocrine system in light of the advances in understanding of receptor signaling and molecular biology.

The great importance of the endocrine system is that it controls the development of the organism and also many vital aspects of the functioning of a developed organism. Hormone signaling is a pervasive mechanism across the life-stages and the various organs and systems. It intervenes in aspects of the development and functioning of the organism as diverse as stimulation or inhibition of growth, regulation of metabolism, induction or suppression of programmed cell death (apoptosis in scientific jargon), activation or inhibition of the immune system, menopause, control of the reproductive cycle, but also wake-sleep cycles, mood swings, etc.

In addressing the difficult issue of endocrine disruption, it is important to be aware of the particular complexity, pervasiveness and sensitivity of the underlying physiological mechanisms. Even if complexity is not rare in toxicology, ecotoxicology and risk assessment in general, endocrinology, which is the basis for understanding the mechanisms of endocrine-related adverse impacts, adds its own intricacies, uncertainties and unknowns to the other aspects of the issue.

The "Endocrine disruptors" Issue The background to the endocrine disruption issue and controversy is characterized by two facts on which there has not been particular disagreement. First, it has been proven that a variety of exogenous natural or man-made chemical substances can interact with the endocrine system. Those are substances that are not produced by the organisms like the hormones: they are exogenous and the organism may be exposed to them through the environment, food and products of different nature (pesticides, cosmetics etc.). EFSA has introduced and defined the concept of "endocrine active substances (EAS) (EFSA 2010):

> An endocrine active substance is any substance that can interact directly or indirectly with the endocrine system, and subsequently result in an effect on the endocrine system, target organs and tissues.

This definition highlights, in addition to the concept of interaction with the endocrine system, the fact that such interaction may take place directly, by the substance acting on receptors, or indirectly through the inducement of production in the body of a derived active substance. Moreover, the interactions considered under this definition lead to "effects" on the endocrine system. Nevertheless, the definition of EAS is not restricted to those substances that can cause adverse effects and includes also the endogenously produced hormones. EAS is therefore a broad category.

It is obviously of particular interest in the discussion on the endocrine disruption to focus on the exogenous, natural or man-made, EAS. Various types of interaction of exogenous EAS with the endocrine system have been observed of which the binding with receptors is the most familiar (EFSA 2013-1).

The second observed fact is that in light of data on the incidence of certain diseases in humans and adverse effects in the wildlife some scientists consider such diseases to be related to exposure to EAS. It is in particular highlighted that many endocrine-related diseases and disorders are on the rise (Åke Bergman et al. 2013):

- Large proportions (up to 40 %) of young men in some countries have low semen quality, which reduces their ability to father children.
- The incidence of genital malformations, such as non-descending testes (cryptorchidisms) and penile malformations (hypospadias), in baby boys has increased over time or levelled off at unfavourably high rates.
- The incidence of adverse pregnancy outcomes, such as preterm birth and low birth weight, has increased in many countries.
- Neuro-behavioural disorders associated with thyroid disruption affect a high proportion of children in some countries and have increased over past decades.
- Global rates of endocrine-related cancers (breast, endometrial, ovarian, prostate, testicular and thyroid) have been increasing over the past 40–50 years.
- There is a trend towards earlier onset of breast development in young girls in all countries where this has been studied. This is a risk factor for breast cancer.
- The prevalence of obesity and type 2 diabetes has dramatically increased worldwide over the last 40 years. WHO estimates that 1.5 billion adults worldwide are overweight or obese and that the number with type 2 diabetes increased from 153 million to 347 million between 1980 and 2008."

Parallel observations are made concerning adverse effects on the wildlife. In concomitance with these two sets of facts, the presence in the environment and in a variety of products of a great number of chemicals with EAS characteristics has been observed. Moreover, in vitro and in-vivo tests have shown that certain EAS may under certain conditions cause endocrine mediated adverse effects on the development and functioning of organisms.

Concerns have emerged on the possible link between exposure to EAS and adverse effects on human health and the environment. These concerns have induced the inclusion in some EU and national legislative instruments of references to and provisions on "endocrine disruptors". We will discuss this term further on. However, the causal link between observed adverse effects and exposure to endocrine active substances remains controversial.

The Concepts of "Mode of Action" and "Mechanism of Action" In toxicological and eco-toxicological studies, once a statistically significant correlation is established between exposure to a chemical and an adverse effect, it is important to explore and understand the mechanism through which the chemical possibly causes the effect in question.

Identifying and understanding the mode or mechanism of action reinforces the evidence supporting a cause-effect link. A mode of action (MoA) describes a functional or anatomical change, at the cellular level, resulting from the exposure of a living organism to a substance. In comparison, a mechanism of action (MOA) describes such changes at the molecular level (EPA 2003). According to this definition, the concept of mode of action applies to any significant effect, not only the adverse effects. In the meaning of this definition, endocrine activity, that is the interaction of a chemical with the endocrine system causing an effect, is a mode of action.

Endocrine Disrupting Substances Although EFSA and other bodies as well as numerous scientists have clarified the meaning of endocrine active substances, indicating that not all the endocrine mediated effects are necessarily adverse effects, the term most frequently used, often without qualification, is that of "endocrine disruptors". Without definitions and clarifications this term may cause some serious misunderstanding. In fact, the common language associates a negative connotation to the verb "to disrupt". The verb "to disrupt" does not just indicate to "cause an effect": it means causing an adverse impact by impeding to proper functioning of an action, a process or a mechanism.

Therefore, the term "endocrine disruptor" if used in its proper sense should apply only to those substances that produce an adverse effect mediated by the endocrine system. The terminological uncertainty related to the use of the words "endocrine disruptors" by commentators, politicians and the media causes two sorts of difficulties in the current debate. First, it is often unclear whether the term is used to refer to the mode of action (in which case the term is not used in its proper meaning and inadvertently implicitly suggests that any interaction of an exogenous substance with the endocrine system producing an effect constitutes an adverse effect (disruption)), or if really it is intended to refer only to those substances capable of causing an endocrine mediated adverse effect.

The second difficulty is the vagueness of the term even when it is used in its proper meaning (substances causing adverse effects) because there is a great variety of endocrine mediated adverse effects: endocrine "disruption" says nothing about the end points involved and therefore the severity and characteristics of the health or environmental damage: it does not inform whether the hazard in presence is carcinogenicity, mutagenicity, repro-toxicity etc. etc.

Overall, the introduction of the term "endocrine disruptor" has been unfortunate. Although the current problems are in substance related to the difficulty of establishing an appropriate, effective and accepted management approach, the debate would have been more transparent if, instead of referring to the hybrid concept of endocrine disruption, mixing mode of action and undefined hazards, the issue had been defined and characterized through the well-established hazard categories and the specific mode of action (endocrine activity).[1]

[1] For instance in the case of carcinogenicity, a distinction is made based on mode of action between the genotoxic and non-genotoxic carcinogens.

EU Legislation on Endocrine Disruptors Different pieces of EU legislation refer to endocrine disruptors.

In the cases of Plant Protection Products (PPP) and Biocidal Products (BP), the relevant EU Regulations provide that a substance having endocrine disrupting properties which may cause adverse effects shall not be authorised unless:

- For PPP:
  - The exposure is negligible, or
  - The substance is necessary to control a serious danger to plant health which cannot be contained by other available means including non-chemical methods (the latter derogation is applicable for a maximum of 5 years)
- For BP:
  - The risks are negligible, in particular where the product is used in closed systems or under other conditions which aim at excluding contact with humans and release into the environment, or
  - The substance is essential to prevent or control serious dangers to human health, animal health or the environment, or
  - Not approving the substance would have disproportionate negative impacts on society when compared with risks.

In both cases, substances with endocrine disrupting properties may not be approved for use in products intended for use by the general public. It is not specified whether the scope of this provision is restricted to those substances which may cause adverse effects. The intention of the legislator was clearly to prevent use of substances with endocrine disrupting properties in PPP and BP, under very restrictive conditions.

The interpretation of these provisions pose a number of problems. There is no definition of "substances having endocrine disrupting properties". Since it is added "which may cause adverse effects", one could be tempted to conclude that substances having endocrine disrupting properties are to be interpreted as equivalent to endocrine active substances, unless the specification "which may cause adverse effects" is to be considered a redundant repetition of a concept implicit in the terms "disrupting properties". If that is the case, the use of endocrine active substances in products intended for use by the general public, should be considered banned independently of an adverse effect (banned on consideration of a mode of action even if not leading to an adverse effect). One could conclude that the legislator has been very precautionary in the case of products intended for the general public in consideration of the difficulty to establish evidence of endocrine-mediated adverse effects, the size of the population potentially exposed and the uncontrolled conditions of use of such products. However, if it is assumed that the wider category of endocrine active substances may pose hazards that are difficult to identify and assess, it is less clear why the protection of workers and the public through the environment is restricted to those substances for which an adverse effect is demonstrated. Obviously, use of the products in question by professionals may be assumed to take place under controlled conditions. Nevertheless, while the provisions on endocrine disrupters which may cause adverse effects are very protective, nothing is said about the use in products for professionals of substances with just endocrine disrupting properties. Should instead be concluded that in all cases the intention is to address those substances showing a potential for endocrine-mediated adverse effects? These questions are also important in relation to the mandate that the EU legislator has conferred upon the Commission to establish scientific criteria for identifying *substances with endocrine disrupting properties*. Moreover, the characterization expressed by the terms "which may cause adverse effects" opens a very wide and open-ended field in terms of definitions, research and testing requirements.

Finally, it should be noted that the PPP Regulation does not refer to risk and by the requirement of negligible exposure practically excludes substances on the basis of their intrinsic characteristics (mode of action and generic hazard). Although, the BP Regulation refers to negligible risk, again based on no exposure, the practical implications may be interpreted to be quite the same as those of PPP.

The already mentioned REACH Regulation too includes provisions on endocrine disruptors. Among the substances that may be subject to an authorization requirement, with a sunset date for their presence on the market, are "substances… having endocrine disrupting properties…for which there is scientific evidence of probable serious effects to human health or the environment which give rise to an equivalent level of concern to those of other substances listed in points (a) to (e) and which are identified on a case-by-case basis…". Points (a) to (e) refer to substances carcinogenic, mutagenic, reprotoxic, PBT (Persistent, Bio-accumulative and Toxic) and vPvB (very Persistent and Very Bio-accumulative). The formulation is complex, poses some interpretative problems and needs a qualitative judgment, case-by-case. The over-abundance of terms and conditions included in this provision may be confusing. One could for instance conclude that certain substances may have endocrine disrupting properties without causing serious effects to human health or the environment: in such a case, substances with endocrine disrupting properties would coincide with the endocrine active substances as defined by EFSA. But is that the intended meaning? And if so, why to refer to the concept of disruption for a substance that is not a-priory considered hazardous? Moreover, in order for a endocrine active substance to fall under the scope of the REACH authorization system, there must be evidence of *probable serious* effects, and such effects must give rise to an *equivalent level of concern* to various well established hazard categories. Although the overall concept is understandable, the qualitative terms used (probable, serious, equivalent) open a wide scope for subjective interpretation, while the regulation does not offer much guidance or criteria.

In conclusion, the provisions related to substances with endocrine disrupting properties in the legal instruments examined regulate such substances on the basis of their mode of action plus unspecified adverse effects on human health or the environment. In the case of REACH, such effects must be probable, serious and equivalent to certain categories of hazards of high concern. Being based on the intrinsic characteristics of the substances considered, these provisions may imply three orders of considerations for their justification (unfortunately not specified in the preamble to the legal texts):

- These substances (under the specified conditions for their intrinsic characteristics) pose an unacceptable risk independently of the level of exposure (for instance if it is assumed that there is no threshold for their adverse effects), or
- It is impossible or not reasonably practicable to ensure that the exposure, and therefore the risk, remains below the level assumed to correspond to the high level of protection chosen, and/or
- The uncertainties and variability in presence on the hazardous properties of the substances, their fate in the environment and/or the levels of exposure are so high that the precautionary principle must be applied and endocrine disrupting substances which may cause adverse effects must be banned based on their intrinsic characteristics (with the exception of those cases where it is possible to demonstrate a negligible exposure).

Therefore, the validity of any of the possible justifications depends on the assessment of the relevant facts and on the scientific knowledge on the possible endocrine-mediated adverse effects. It is against this background that an intense a bitter debate has developed, both on the scientific aspects and on the management approaches and implications, these various aspects being closely interrelated. The debate has focused around the Commission activities aimed at defining criteria for identifying substances having endocrine disrupting properties.

The Search for Criteria and Definitions A definition of endocrine disrupting substance was established by the WHO and is widely referred to, even if it has so far not been incorporated into the relevant regulations. WHO makes a distinction between an endocrine disruptor and a potential endocrine disruptor (WHO/IPCS 2002):

- An endocrine disruptor is an exogenous substance or mixture that alters function(s) of the endocrine system and consequently causes adverse health effects in an intact organism, or its progeny, or (sub)populations.
- A potential endocrine disruptor is an exogenous substance or mixture that possesses properties that might be expected to lead to endocrine disruption in an intact organism, or its progeny, or (sub)populations.

This definition includes the concepts of mode of action, adversity of health effects and causality between the mode of action and the adverse effect.

Adversity

As already noted, the concept of adversity is potentially very wide. It is not clarified in the above-mentioned definition. And there is no reference to any specific hazard. Therefore, it may be difficult to find a consensus on the frontier between effects and adverse effects. The German Federal Institute for Risk Assessment has proposed in 2009 the following definition of adversity (BfR 2009):

> A change in morphology, physiology, growth, reproduction development or lifespan of an organism which results in impairment of functional capacity or impairment of capacity to compensate for additional stress or increased susceptibility to the harmful effects of other environmental influences.

Although this is a conceptually comprehensive definition, it involves obvious application difficulties, notably in defining the screening and testing requirements and modalities (for example: which tests and criteria would allow deciding on the "*increased susceptibility to the harmful effects of other environmental influences*"?).

Causality

Due to the complexity of the endocrine system and the variety of effects and adverse effects that may arise, it may be difficult to provide proof of a causal link between exposure to a certain substance and an observed outcome. We have seen that the REACH Regulation requires evidence of "probable serious effects", while there is no such qualification in the PPP and BP regulations. WHO has modulated its definitions making a distinction between "endocrine disruptors" and "potential endocrine disruptors". There is concern among certain experts and stakeholders that too strict weight of evidence criteria could exclude from the scope of the legislation many potentially hazardous substances.

The Controversy Around the Definition and Management of the Endocrine Disrupting Substances Although the EU legislator, the European Parliament and the Council, have established the regulatory approach applicable to endocrine disrupting substances since several years, the controversy on both the scientific and the management aspects of the issue have reached unprecedented levels around the definition by the Commission of criteria for the identification of such substance, which was due by 2013. Unfortunately, in addition to the intrinsic difficulties of the subject, the argument has been characterized by a worrying confusion of roles, with scientists taking strong public positions on management and regulatory aspects. The bitter argument between different groups of scientists, taking side with opposing policy stances or economic interests, has damaged the credibility and utility of scientific advice on this important matter, pushing the European Commission to embark in a very comprehensive re-assessment of all the aspects and implications of the subject within the framework of an impact assessment. The Commission has also organized in 2015 a series of dialogues involving scientists and stakeholders, trying to clarify the points of disagreement and to reconcile as far as possible the divergent positions.

On substance:

- Some scientists consider that the endocrine system is so complex that it is difficult to identify all the health-significant effects or to exclude potential adverse effects in case of exposure to endocrine active substances. Moreover, they consider that the tests prescribed by the applicable legislation are not adequate to cover all the possible end-points. Certain effects show only in the long-term and are dependent on the period of exposure (pre-natal or during critical windows of developments). Finally, they express doubts about the existence of dose-thresholds below which endocrine active substances could be expected not to cause adverse effects.

– Another groups of scientists have opposite views, practically on all points. They think that there is no great difference between endocrine-mediated effects and effects deriving from other modes of action. For them, the classical risk assessment approach is applicable independently of the mode of action. The adverse effects of certain endocrine active substances overlap with some well-established hazard categories (carcinogenicity, mutagenicity, repro-toxicity). They consider that progress has been made in the development of appropriate testing approaches. In their opinion, there is no reason why endocrine disrupting substances should show no threshold. Finally, the different potency of various substances should be taken into account in the assessment of health risks.

Behind such disagreement on scientific aspects, there are divergent views and interests on the way in which endocrine disrupting substances should be regulated. Although the EU legislator has already decided to regulate on the base of the mode of action and the generic concept of "adversity" (however with provisions which may pose some difficult interpretation and implementation problems as we have seen), industrial interests supported by part of the scientific community push for a revision of the approach, defending the applicability of the classical risk assessment concepts. The debate is however complicated by the terminological confusion discussed above, the absence of a clearly expressed rationale behind the current legislation and the fact that substantive regulatory issues are discussed within a process that is in fact limited to establishing by a delegated act of the Commission criteria for identifying endocrine disruptors.

An Unfortunate Confusion of Roles The debate on endocrine disruptors has shown in 2014 a worrying confusion of roles, with a group of scientists launching a strong attack on the European Commission based on non-scientific considerations. The "casus belli" was offered by an internal document of one of the Commission services, unduly leaked, containing a draft recommendation on "defining criteria for endocrine disruptors". That not-official, internal document did not represent a position endorsed by the Commission. Moreover, the proposed definitions of endocrine disruptor and suspected endocrine disruptors were those established by the WHO/IPCS, well known and widely accepted, limiting the identification of endocrine disruptors to those substances for which there is evidence of an endocrine mediated adverse effect, with a clear distinction between endocrine activity and endocrine disruption. The envisaged draft recommendation, by its nature non-binding, did not touch upon risk management issues.

In a very surprising move, by the end of 2013 a group of chief-editors of scientific journals published an editorial with the title "*Scientifically unfounded precaution drives European Commission's recommendations on EDC regulation, while defying common sense, well-established science and risk assessment principles*" (Dietrich et al. 2013). The editorial was as violent, nearly offensive in its language, as pointless and full of factual errors. It started by expressing concern on "the imminent decision by the European Commission to enforce a regulatory framework for so-called endocrine disrupting chemicals". Apparently the editors had not noticed that a regulatory framework, decided upon by the European Parliament and the Council, was already in place, that a recommendation is not a regulatory instrument and that recommending criteria for the identification of endocrine disrupting substances does not determine, modify or "enforce" the applicable regulatory framework. It went on accusing the Commission to ignore "all well-established and taught principles of pharmacology and toxicology". Given that the recommendation was just about criteria for endocrine disruptors, taking on board both the definitions of WHO/IPCS, endorsed by dozens of scientists, and the distinction introduced by EFSA between endocrine activity and endocrine disruption, the criticism sounded quite unfounded. The editorial explained the difference between mode of action and hazard, and insisted on the fact that not all effects could be considered adverse effects. All these considerations were however pointless, because the draft recommendation was based exactly on the same concepts (it was proposed to consider as endocrine disruptors only those substances for which there is evidence of a endocrine-mediated adverse effect). In a crescendo of allegations the editorial went on attributing to the Commission the intention to foresee

> a priori regulation of agents that may show presumably endocrine-mediated effects in some experimental system...and under the a priori default assumption of no threshold.

The draft recommendation was proposing exactly the opposite: to consider as endocrine disruptors only the substances showing endocrine-mediated adverse effects. It did not pronounce on the existence or not of thresholds, something that is checked in the toxicological studies and is relevant for characterizing hazards and risks and for management decisions, not for identifying the intrinsic hazardous properties of a substance. Moreover, the editorial criticized, based on obvious but generic considerations on the impossibility to demonstrate the absence of effects, a provision of the draft recommendation stating that adverse effects demonstrated in animal tests should be assumed to be significant for humans unless there is evidence on the contrary, a quite common assumption without which most of the experimental toxicology based on animal models would be worthless. Then, the signatories, abandoning any "scientific" reserve, expressed concern for

> regulation that affects human activities and is contrary to the social contract.

They felt the duty to

> resist and counteract any effort that undermine the core of science and its continuing promise for the betterment of the human condition and of the planet.

The content and tone of such statements (related to an internal non-official draft document emanating from just one service of the Commission, unduly leaked, concerning a project for a non-binding recommendation, clearly grossly misinterpreted!) was surprising and inappropriate. A clearer position was expressed by a group of scientists in a letter to the then "Chief Scientific Advisor" to the President of the European Commission, which was publishes as an annex to the editorial mentioned above. The signatories were clearly aware of the regulatory framework in place and the fact that the EU legislator (the EU Parliament and the Council) had opted for an approach based on consideration of mode of action and adversity, instead of the classical risk-based approach. They were also aware that in the absence of criteria for identifying endocrine disruptors, the adopted legislation could not be effective in practice for managing such substances. Their letter was therefore a more transparent and coherent plea to block the adoption by the Commission of endocrine disruptors criteria. The argument were more balanced and focused than those of the editorial. Nevertheless, it was worrying to see scientists pronouncing on a risk management approach decided by the European Parliament and the Council, trying to interfere in a quite oblique way with its implementation, particularly considering the lack of consensus within the scientific community on the scientific arguments invoked.

The editorial and the letter of scientists called for a reaction by other scientists and stakeholders. In particular, a group of scientists published in August 2013 an article on "Environmental Health" (Åke Bergman 2013) with the title "Science and policy on endocrine disrupters must not be mixed: a reply to a "common sense" intervention by toxicology journal editors" strongly disagreeing with and expressing concern on the statements of the editorial, that the signatories considered "inaccurate and factually incorrect and calling for "a better founded scientific debate which may help to overcome a polarisation of views detrimental to reaching a consensus about scientific foundations for endocrine disrupter regulation in the EU". The Article referred in particular to the views that had been expressed by the 129 signatories of the "Berlaymont Declaration" on endocrine disrupters[2] and by the Collegium Ramazzini. The arguments of the editorial were criticized in detail based on scientific considerations.

The distance of positions and bitterness of the argument expressed by scientists on the issue of endocrine disrupting chemicals show how sometimes science could be unable to support policy making with consistent, neutral, evidence-based advice. The initiative of a group of scientists to expose the divergence of positions between

[2] *The Berlaymont Declaration on endocrine disruptors* 2013, http://www.ipcp.ch/IPCP_Berlaymont.html

experts, criticising the Commission in very rude terms, was anyway successful in creating confusion on the real degree of scientific disagreement, blocking progress towards a timely adoption of scientific criteria for endocrine disruptors, pushing the Commission to embark in a complex impact assessment, missing the deadline of 13 December 2013 set in the legislation on biocidal products for the adoption of criteria by a delegated act, and finally resulting in a decision of the General Court of the Union in December 2015 concluding that the Commission had failed to fulfil its obligations.

At the time of writing this book, the endocrine story is not yet concluded. We say more on the lessons from this case in Chap. 7.

4.5.12 The Assessment of Nanomaterials

A new issue for health risk assessment of chemicals has appeared with the development of nanotechnologies. The basis of nanotechnologies is the exploitation of the properties of matter at the nanoscale, corresponding to a size on the order of one billionth metre. Certain substances, when in a form (nanoparticles or nanostructures) allowing interaction at that level of size, show physical and chemical properties different from those of the usual "bulk" form. Nanomaterial are researched, produced and used precisely when they show "new", different useful properties. However, their novel properties may also include different biological properties or behaviour in biological systems. That poses the problem of the validity of toxicological profiles of substances determined for their "bulk" form, when they are used at the nanoscale. Reviews of the potential risks of nanomaterials have concluded that "nano" is not synonym of toxic, and that the risk assessment methods developed for chemicals are adequate for the risk assessment of nanomaterials, but that specific data on the toxicity of such material, including their toxico-kinetics and toxico-dynamics may be necessary.

Risk managers should be attentive to the need to obtain specific data on the toxicity of nanomaterials. Risk assessors critically review the validity of extrapolations from bulk to nano-form of a substance. The EU risk assessment bodies have delivered several opinions clarifying the specific aspects of the application of risk assessment methods to nanomaterials.

4.5.13 Interpreting Scientific Advice on Risks

Scientific risk assessment of chemicals is a well-developed decision-aid tool for supporting rational, effective and efficient decisions. However, decision makers may find it difficult to interpret and use advice on risks of chemicals provided by scientists. Several characteristics of the risk assessment methods and practices may complicate the task of decision makers in that respect.

First, health risk assessment of chemicals involves the application of a variety of disciplines including, among others, medicine, biology, toxicology, pharmacology, statistics, and of a set of methods and criteria specific to risk assessment. Understanding the process of the assessment, the scientific reasoning, and the terminology used by assessors to express findings and results may be quite difficult for those decision makers who have no scientific background and are not familiar with the assessment methods.

Second, the very nature of risk assessment is often misunderstood. Policy makers often expect definitive, univocal and unchallengeable answers from risk assessors; they may not always realise that "scientific" risk assessment is indeed a science-based process, but, as we have already noted, it is not "pure science". Outside the specialist circles, there is little awareness of the limitations often imposed by the state of knowledge and/or the scarcity of data. Knowledge and data that are needed for a full risk assessment are sometimes simply not available. Moreover, although the broad methods and criteria for the assessment of health risks caused by the exposure to dangerous chemicals are well established and internationally recognised, there is no single, rigorously standardised or formally codified approach. There are still important aspects for which agreed methods are missing and further research and development work is needed. Risk assessors bring in their professional judgement in weighing evidence, selecting data, making assumptions. The results of a risk assessment may often require qualification and interpretation.

Dissenting opinions of qualified experts are often heard, criticising mainstream risk assessment conclusions on important and controversial issues, like for examples GMOs or endocrine disruptors. There are many possible reasons for diverging opinions: different criteria might be used by the assessors for selecting and weighing evidence and data; differences may arise from data interpretation; assessors may disagree on the models or exposure scenarios to be used etc.

We have shown that scientific advice on risks is most often affected by some degree of subjectivity and approximation. Sometimes, the extent, seriousness or even the very existence of a risk is controversial among different scientists or scientific bodies. Even some basic questions remain unresolved, like for example the way to deal with "endocrine disrupting substances. In spite of the importance of uncertainties in risk assessment, there is no generally recognised framework, applicable across the various risk-sectors, for the treatment and the expression of uncertainties. Even the quantification and expression of risks are not harmonised as a report on the practices of the Scientific Committees of the European Commission has shown (UK Central Science Laboratory 2007). And risk assessment results are often expressed in technical terms, like for example margins of safety, whose implications for risk management are not straightforward.

A better understanding by decision makers and stakeholders of how risk assessment works is a pre-condition for improving the role and utility of scientific risk assessment in the decision and governance processes. All actors and interested parties should be clear about what risk assessors can and cannot deliver, and should be aware of the potential, but also of the limitations of the risk assessment methods.

Ensuring a better understanding of risk assessment methods beyond the specialist circles would facilitate a productive dialogue and a more systematic collaboration between scientists, risk managers and decision makers. We say more on this subject in Chap. 8.

4.6 Assessment of Microbiological Risks

4.6.1 Differences Between Chemical and Microbiological Risk Assessment

The exposure to bacteria, viruses, fungi and other microorganisms from various sources and through different routes may involve significant health risks depending on a wide range of conditions. Microbiological risks are an important risk category in relation to EU policy and regulatory responsibilities notably in the area of food safety, the safety of medical devices, cosmetics and other consumer products as well as for certain aspects of the EU public health policy. The EU risk assessment bodies assess microbiological risks in order to provide a basis for regulatory decisions or for supporting EU-level action in case of crisis following accidental contaminations or outbreaks of infectious diseases.

Assessment methodologies for microbiological risks are conceptually aligned with the approach applied to chemicals. However some important aspects are specific to microbiological risks, with significant methodological implications.

Characteristics of Contamination by Microorganisms Differently from chemicals, which may cause human or environmental exposure both as contaminants and as intentionally added substances, microorganisms are normally present only as non-intentional contaminants. As such, their presence, type, characteristics and quantity in a specific product or environmental media may be particularly difficult to determine. Moreover, the number and characteristics of microorganisms are time-dependent because microorganisms reproduce themselves and evolve rapidly. Proliferation and changes depend on several factors, including in particular the vector, the environmental conditions, the treatment applied etc. Microorganism proliferation is a peculiar aspect and represents a substantial difference compared to chemicals, which can concentrate, bio-accumulate, transform in the environment and give rise to metabolites in the organism, but cannot reproduce themselves.

Detection of Microorganisms Detection of very low concentrations of pathogens with the available methods is difficult. While the methods available for detecting chemicals are sufficiently sensitive to provide information on the presence of contaminants at the concentration levels that are significant for human health, the same cannot be said for microorganisms that can cause infection even starting from a very low presence of pathogens. For a same type of microorganism, there is often a need to make a distinction between different strains. Things are complicated by the rapid evolution of microbial *genome*, rapid mutations and horizontal gene transfer that can modify drastically the pathogenic potential and the virulence of microorganisms.

Different Susceptibility to Microorganism Pathogenicity Finally, the pathogenic effects may greatly differ depending on the immunological status and other differences in susceptibility of the exposed population groups. As a consequence of all that, microbiological risk assessment is characterised by its own specific complexities, variability and uncertainties.

Since the approach of microbiological risk assessment is conceptually similar to that previously presented for chemicals, the objective here is to briefly highlight only those aspects, issues and problems that characterise specifically the case of microorganisms, with special focus on food safety.

As in the case of chemicals, microbiological risk assessment includes the four steps of hazard identification, hazard characterisation, exposure assessment and risk characterisation.

4.6.2 Biological Hazard Identification

For regulatory purposes, assessors are often asked to assess the potential risk of contamination of a given (food) product by a certain type of microorganism. In such cases, the hazard identification consists of the determination of the potential for pathogenic effects on humans of the microorganisms in question, transmitted by the product considered, under the conditions applicable to the case.

Sometimes the nature of the microbiological contamination is unspecified. In such a case, assessors must first determine the type of microorganisms contaminating the product considered and then identify their pathogenic characteristics.

A still different case is the identification of the causal link between a public health problem, like the increased incidence of a given pathology, and a possible exposure to a certain type of microbiological contamination.

In all the cases above, hazard identification requires human or animal data on the pathogenic characteristics of the microorganisms considered.

4.6.3 Biological Hazard Characterisation

Hazard characterisation is aimed at determining the (quantitative, if possible) relationship between the level of exposure, the dose, and the probability and severity of the adverse health effect. The relevant aspects of microbiological hazard assessment are: the characteristic of the pathogen, the host and the vector (food product) matrix. The interaction between these elements may be very complex.

Many factors may alter the infectivity, pathogenicity, virulence and type of health damage of a microorganism. The variability of the possible health effects depends to a great extent on the host (age, sex, race, health status, immune status etc.). In addition, the matrix (e.g. food type) may significantly influence the pathogenic potential of a microorganism.

It is generally assumed that the dose-effect relationship for microbiological hazards has no threshold. The sources of data for establishing such relationships are human volunteer and animal studies. Due to ethical, cost and practical limitations, only data related to high dose-levels are in general available. As in the case of chemicals, it is therefore necessary to extrapolate from such data, down to the levels representative of human exposure. Depending on the mathematical model adopted, very different results may be obtained. Efforts are made by the assessors to select models that make biological sense.

Dose-effects studies have limitations similar to those highlighted for chemical hazards. Nevertheless, in the case of microbiological risks the higher variability associated with the host characteristics, introduces even higher uncertainties. Dose-response ranges in the case of exposure to pathogens may vary by several orders of magnitude between the various strains of the same species.

4.6.4 *Exposure Assessment*

In exposure assessment, the routes through which individuals may get in contact with the pathogen must be determined. In the case of food, the sources, mechanisms and fate of the contamination must be explored. The impact of processing and storage must be taken into account. Differently from chemicals, exposure can also come through secondary transmission, between humans or through animals.

4.6.5 *Risk Characterisation*

As in the case of chemicals, risk characterisation is aimed at providing a qualitative or, when possible, quantitative estimate of the probability and severity of the possible adverse health effects of exposure to the microorganisms considered, under the specific conditions and circumstances applicable. The process consists in combining the data and information on the hazard of the pathogens, the exposure and the exposed population. The resulting adverse effects may be expressed as incidence of pathologies, deaths and their economic consequence in terms of healthcare costs, lost revenue, lost working days etc. Considerations on the limits and uncertainties of the final risk estimate are not different from those applicable to the case of chemicals, account taken obviously of the specific complexities and sources of variability and uncertainty for the assessment of microbiological risks.

4.7 Ecological Risk Assessment

4.7.1 The Objectives of Ecological Risk Assessment

Ecological risk assessment is aimed at identifying and evaluating the risk for ecosystems and for the environment in general, caused by stressors that may be chemical, physical or biological. Such risks may be related to the use of products, to the introduction of technologies, to projects and installations of various nature (infrastructures, industrial plants etc.), or to land use changes. Examples are use of fertilisers and pesticides in agriculture, emission of pollutants from power plants, deforestation, and construction of dams. Ecological risk assessment addresses possible damages to particular species of fauna and/or flora.

The typical negative effect considered is the deterioration of the ability of an ecosystem to sustain life of the species that depend on its quality. In environmental risk assessment, the negative effect may be, in particular, the deterioration of an environmental media in a certain region or area, considered per-se, independently of the services delivered by the environmental media at stake. The aesthetic characteristics of landscape may also be considered a good to be protected.

The basic concepts of ecological risk assessment are very similar to those of health risk assessment. The starting step is the identification of the possible negative effects of the stressor considered and their characterisation. The exposure of the ecosystem or other environmental targets is then assessed. And the final step is the characterisation of the risk. Nevertheless, ecological risk assessment differs from health risk assessment on several important aspects.

4.7.2 Differences Between Health and Ecological Risk Assessment

Differently from the narrow case of human exposure, a great variety of ecological and environmental targets may need to be considered in ecological risk assessment that show significantly different responses to a given stressor. Since complete data is rarely available, inter-species extrapolation is often the only way to make a comprehensive assessment.

Problem and Metric Definition In ecological risk assessment, problem definition is of a particular importance in order to focus the assessment on the aspects that are considered important by risk managers. Many ecological impacts may be expressed and sometimes quantified in biological terms. However, in order to express them as negative effects, they have to be translated into loss of utility from the human point of view.

Exposure Assessment Exposure assessment may be particularly complex when the stressor has to go along a multi-stage path (which may include transportation, diffusions, reactions and transformation in the case of chemicals) before reaching

Table 4.3 The air pollutants transfer across media

The source of air emissions
The transportation and transformation in the atmosphere
Deposition
Uptake by animals, plants, microorganisms
Bioaccumulation
Direct and indirect exposure of man

the ecological target at stake. An example is provided by air pollutants emitted by an industrial facility that may travel long range, across various environmental media, as shown in Table 4.3.

4.7.3 Phases of Ecological Risk Assessment

Ecological risk assessment includes three phases, well described in the US-EPA Guidelines for Ecological Risk Assessment (U.S. EPA 1998). According to EPA guidelines, the process of an ecological risk assessment begins with planning and research. It includes:

- Phase 1 – Problem formulation: Information is gathered to help determine what, in terms of plants and animals, is at risk and what needs to be protected.
- Phase 2 – Analysis: This is the determination of what plants and animals are exposed and to what degree they are exposed, and if that level of exposure is likely or not to cause harmful ecological effects.
- Phase 3 – Risk Characterisation: Risk characterisation includes two major components: risk estimation and risk description. "Risk estimation" combines exposure profiles and exposure-effects. "Risk description" provides information important for interpreting the risk results and identifies a level for harmful effects on the plants and animals of concern.

In the assessment of ecological risks, the objective is often to determine the predicted environmental concentration (PEC) and to compare it with the predicted no-effect concentrations (PNEC) for any given pollutant and for the exposed environmental media and ecological target, in the region and location considered. This approach is similar to that of "safety assurance" for human health, based on consideration of a NOAEL and the exposure level.

4.8 Risk Assessment of Food and Feed from Genetically Modified Plants

The health and environmental safety of genetically modified organisms (GMO) is one of the most controversial risk-issues that the European Institutions have encountered in recent times. Modifying the genetic profile of organisms raises philosophical, religious or ethical concerns that are rooted into our deep beliefs about our relationships with nature. There is often a feeling that manipulating the forms of life resulting from billions of years of evolution may lead to consequences that science is not in a position to foresee. In addition, certain technological developments in the area of biotechnology have encountered resistance for economic reasons, the resistance of EU consumers to accept GMO-derived products and the political difficulty to endorse measures based on mainstream scientific advice alone in presence of vocal minority opinions. Risk assessment of genetically modified organisms has made great progress in recent years, but there is not yet a full consensus within the scientific community on the possibility to exclude any significant health or environmental risk for certain applications of biotechnology.

An exhaustive analysis of the safety assessment and management of biotechnologies goes beyond the scope of this book. We focus here on the broad lines of the risk assessment approach for the food and feed from genetically modifies plants. The scope of this short introduction to plant GMOs is limited to the assessment of health effects. The objective is to show what in such an approach is similar to the case of chemicals and what are instead the more specific aspects and concepts for the genetically modifies organisms considered. It should be noted that the potential environmental impacts of genetically modified plant cultivation are assessed with specific methods that are not introduced here. Much of the debate on "green" GMOs is related to their environmental impact, but consumer rejection of GMOs is mainly focused on their presence in the food chain. We introduce here some concepts and definitions that help understanding the wider debate on GMOs.

4.8.1 Genetically Modified Organisms

According to EU legislation, genetically modified organisms are defined as (EU 2001)

> genetically modified organism (GMO)' means an organism, with the exception of human beings, in which the genetic material has been altered in a way that does not occur naturally by mating and/or natural recombination.

The EU legislation refers also to a non-exhaustive list of genetic modification techniques

Techniques of genetic modification …are inter alia:

1. recombinant nucleic acid techniques involving the formation of new combinations of genetic material by the insertion of nucleic acid molecules produced by whatever means outside an organism, into any virus, bacterial plasmid or other vector system and their incorporation into a host organism in which they do not naturally occur but in which they are capable of continued propagation;
2. techniques involving the direct introduction into an organism of heritable material prepared outside the organism including micro-injection, macro-injection and micro-encapsulation;
3. cell fusion (including protoplast fusion) or hybridisation techniques where live cells with new combinations of heritable genetic material are formed through the fusion of two or more cells by means of methods that do not occur naturally.

4.8.2 Risk Assessment of Food and Feed from Genetically Modified Plants

Steps of Risk Assessment Health risk assessment of food/feed from genetically modified plants is structured along the same steps of the assessment of chemicals. *Hazard identification* aims to identify the biological, chemical or physical agents in modified plants and derived products that have the intrinsic potential for causing adverse health effects. In the specific case of food and feed from genetically modified plants, hazard identification is based on the identification of differences with reference to a relevant "comparator" organism or product as explained later in this paragraph. *Hazard characterisation* aims to establish the qualitative or quantitative relationship between the presence of the dangerous agent and the adverse health effect, when possible in the form of a dose response relationship. With *exposure assessment*, the human (or animal) intake of food (or feed) is estimated, as well as exposure to plant constituents like pollen where relevant. Exposure is characterised in relation to its level, frequency, duration etc. Any significant intake difference compared to the comparable traditional products is noted. As in the case of chemicals, *risk characterisation* brings together the information on hazard and exposure in order to provide a quantitative or qualitative estimate of the severity and probability of occurrence for the potential adverse health effects identified.

The Comparative Approach A specific feature of the risk assessment of food and feed from genetically modified plants is the so-called "comparative approach". The basic assumption is that if the modified plant does not present any significant difference (is *substantially equivalent*) compared to traditionally cultivated crops for which there is history of safe use for human or animal consumption, it can be concluded that the genetically modified plant does not pose a health risks. In such cases there is no need for specific toxicity studies. Such an approach is applicable when a suitable "comparator" is available. The comparison takes into account the natural variability of the comparator crop. The comparative assessment includes a comprehensive molecular characterisation of the modified plant and a comparative study of the modified plant and the comparator, notably their compositional, phenotypic and agronomic characteristics. First, the modified plant and its comparator are tested for the differences that may have a potential to cause adverse effects. Then an equivalence test is applied in order to determine whether the differences identified for the modified plant remain within the range of the natural variability for the comparator.

Toxicological Assessment The usual in vitro or in vivo toxicity tests already described in the section devoted to the chemical risks are also applied to genetically modified plants. In particular, toxicity is studied in vivo by performing tests on animals and applying safety or uncertainty factors to the results in order to determine "safe" exposure levels for humans. Toxicological studies address both the intended and the non-intended effects of the genetic modification, in order to determine whether they can cause adverse effects in humans (or animals), and to characterise such effects. Are considered in particular newly expressed proteins, new constituents, changes in the composition and level of endogenous constituents beyond the natural variability.

Allergenicity Assessment Food allergies may have very adverse health effects and even cause death. Allergy is an abnormal reaction of the immune system to the contact with certain substances. It depends on genetic factors and proceeds in two steps, the first being the sensitisation of the subject to the allergenic factor and the second the contact with the factor leading to the clinic symptoms. In the case of food, the study focuses on proteins that may show allergenic properties. For genetic modified plants, the attention is on the newly expressed proteins. Study of allergenicity is complicated by the fact that the adverse effect is not related to an intrinsic characteristic of the factor but depends on the reaction of the individual organism and its interaction with the factor. Therefore a case-by-case approach is applied, integrating the results from a battery of tests and assessing the results in a "weight of evidence" perspective.

Nutritional Assessment The objective of the nutritional assessment is to verify whether the genetic modification has any adverse impact on the nutritional properties of the derived food or feed. For food, are studied in particular the levels of nutrients and anti-nutrients, their bioavailability and any significant induced variation in the anticipated dietary intake. Variations are determined by comparison with the appropriate "comparator", account taken of its natural variability.

Exposure Assessment Exposure assessment is focused on the determination of the expected intake of food (or feed) derived from the genetically modified plant. Not differently from the case of chemicals in food, the objective is not just to estimate an average intake, but to study the expected distribution, notably the maximum level of intake that can be expected, in order to identify the population or groups at higher risk.

Risk Characterisation Risk characterisation should in particular allow to conclude whether the consumption of food or feed derived from the modified plants is as safe as the appropriate "comparator" plant and whether there is comparatively any adverse impact on the nutritional properties of the food/feed in question.

4.9 Assessment of Radiation Risks

4.9.1 Introduction

For decades, since the beginnings of the nuclear energy industry, the potential risks related to exposure to ionising radiation emitted by radioactive substances have been a cause for public concern. Risk of nuclear industry is still a controversial subject for debate. Nowadays, the potential risks of exposure to *non-ionising* radiation are also a cause for public concern in our technological societies, characterized by intensive use of electronic devices for an increasing variety of functions, like for example communication or medical applications, that are sources of human exposure to low-energy electromagnetic waves.

The EU is in the front line of radiation risk regulation in Europe. For certain types of radiation, the most relevant sources of exposure are products, devices or technologies that are regulated on the EU-level. Moreover, the EU is setting and adapting to the progress of scientific knowledge rules and standards for the protection of workers professionally exposed to radiation as well as rules and standards for preventing or limiting the exposure of the general population. Regulating radiation risks requires a good understanding of the nature of the different types of radiation, their potential risks, the way in which radiation risks are assessed as well as the specific rationale and criteria for radiation risk management. The aim of this section is to provide an introduction to such subjects.

Although the risk analysis paradigm introduced for chemicals also applies to radiation, radiation risks are in fact quite different from chemical risks and their proper management requires at least a basic understanding of the physical nature of radiation.

Therefore, we start our introduction to radiation risk assessment and management with the necessary basic information on the nature of radiation and the categorization of the different types of radiation.

4.9.2 Types and Characteristics of Radiation

Different Types of "radiation" The generic term "radiation" is commonly used to designate different physical entities which, for the aims of risk analysis, are all characterised by the fact that they carry and release energy.

The risk-relevant aspect that the different types of radiation have in common is therefore their capability to release energy into matter, notably the body tissues and organs exposed.

Radiation may consist of electromagnetic waves or sub-atomic particles. Although modern physics associates waves to particles and, in turn, attributes to electromagnetic radiation a dual, wave-particle nature, an important distinction needs to be done between the two types of radiation: differently from sub-atomic particles, electromagnetic radiation does not possess a mass.

Electromagnetic Radiation Electromagnetic radiation is a periodic oscillation of the electromagnetic field. It is a wave that propagates without need for a sustaining medium.

Electromagnetic radiation can travel in the void or, depending on its frequency, across certain media. When meeting charged particles, an electromagnetic field exerts forces on them. As all matter is composed of charged particles, electromagnetic radiation has the potential to release energy into matter by acting on its constituent molecules and atoms (Fig. 4.3).

Electromagnetic radiation is characterised by its frequency (or its wavelength) and its intensity. The frequency is the number of oscillations per second, which is the number of times in a second that the electrical and magnetic fields return to a given value in a certain position of space, and is measured in Hertz (Hz), with 1 Hz equal to one oscillation per second. The wavelength is the distance between two points where the wave assumes the same value. The relation between the frequency f and the wavelength λ is:

$$f = v / \lambda$$

where v is the speed of the wave. The energy level of an electromagnetic radiation is proportional to its frequency:

$$E = h \times f \ (h \text{ times } f)$$

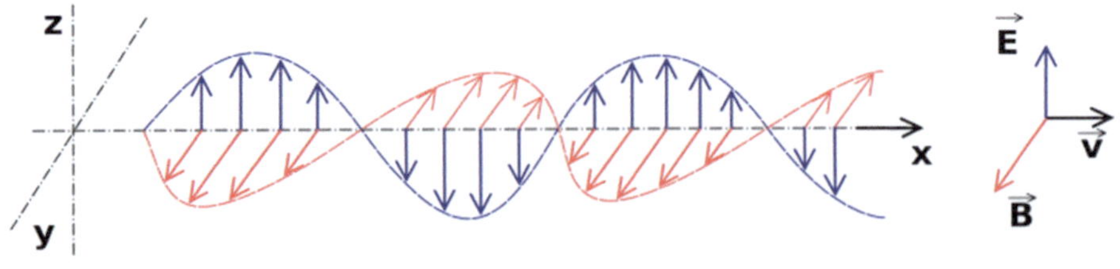

Fig. 4.3 Characterisation of electromagnetic waves (Source: SuperManu, under GNU FDL)

where h is a numerical constant (the Planck constant). Therefore, the energy that electromagnetic radiation can release and its capability of interacting with matter depend on its frequency.

For instance, in order to "kick off" an electron from an atom, the electromagnetic radiation must have at least the frequency corresponding to the energy required to break the "link" that keeps the electron attached to the atom. In the corpuscular model of electromagnetic radiation the energy is transported by particles without mass, the photons, each with an energy expressed by the product of the Planck constant h times the frequency f.

The frequency of the radiation is most important in determining the type of interaction of electromagnetic waves with matter. Such interaction depends on the capability of the individual photons to produce a given effect, for instance the breaking of molecules of genes in the cells making up living matter, which may lead to gene damage.

The number of photons per second reaching a given target determines the intensity of the radiation to which the target is exposed. The total energy transported by the radiation investing the target is therefore proportional to both the radiation frequency and its intensity.

In conclusion, the type of effect of electromagnetic radiation depends only on its frequency, not on its intensity. The size of a given effect depends on both the intensity and the frequency of the radiation (how much energy each photon possesses and how many photons per unit of time invest the target).

The Health Relevance of Electromagnetic Radiation In light of what we have explained, it is clear that the health relevance of electromagnetic radiation depends in the first instance on its frequency. For the needs of risk analysis, it is important to classify electromagnetic radiation on the basis of frequency. In fact, the radio-waves that allow us to communicate or listen to the news, the heat radiated from the fireplace that warm us, the sunlight or artificial light that allows us to see, the ultraviolets against which we protect ourselves at the beach with sunscreens, the x-rays by which the dentist checks whether we have a tooth problem, the gamma-rays that scare us when we hear of nuclear waste risks, are all the same thing physically, but, as they differ by their frequency, show very different properties, including the existence and type of health hazards.

As shown in Fig. 4.4, the frequency of electromagnetic radiation may vary by many orders of magnitude:

Radiation Consisting of Particles The term radiation is also used for certain charged sub-atomic particles that, differently from electromagnetic radiation, possess a mass.

The atom is composed of protons, neutrons and electrons (Fig. 4.5). Protons and neutrons are constituents of atomic nuclei. Protons have a positive electric charge (equivalent but opposite to that of an electron), while neutrons have no charge. Protons and neutrons have approximately the same mass.

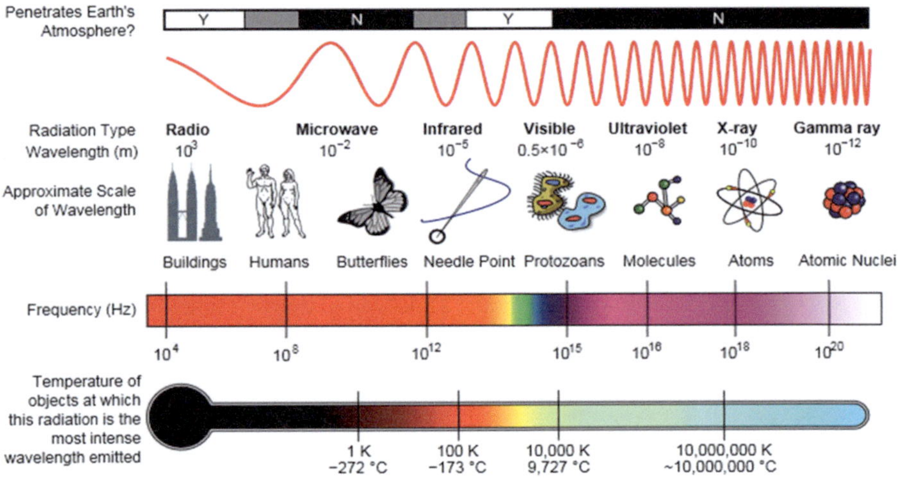

Fig. 4.4 The electromagnetic spectrum (Source: US NASA (public domain))

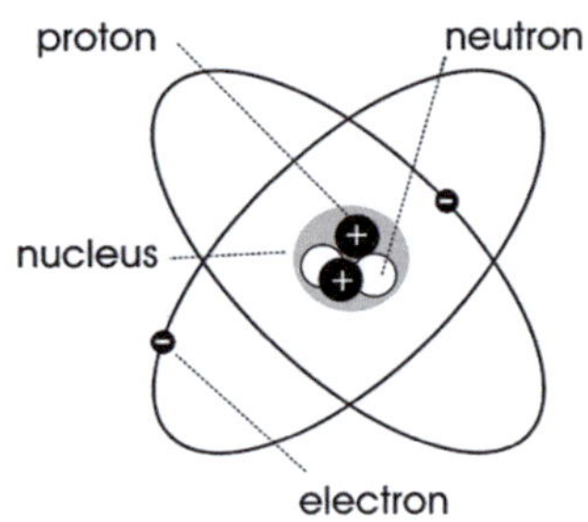

Fig. 4.5 The atom (Source: http://www.wpclipart.com, Public Domain)

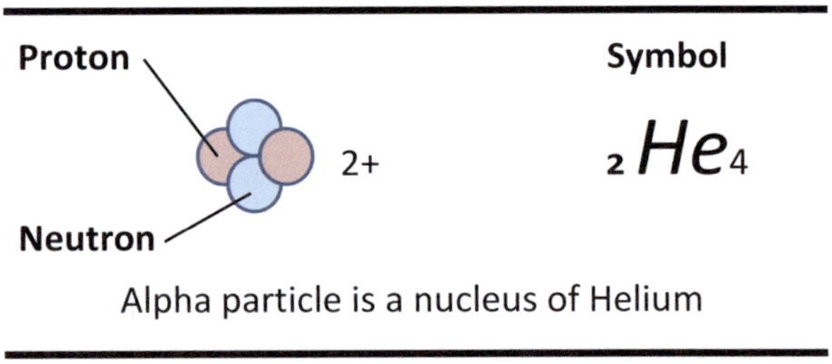

Fig. 4.6 The alpha particle

Alpha particles are constituted by two protons and two neutrons bound together. An Alpha particle is a nucleus of the atom Helium, without its two electrons (a Helium ion) as shown in Fig. 4.6 (*Source*: http://www.ooshutup.com).

Beta particles are free electrons. They have a very small mass and a negative charge. *Neutrons*, the non-charged constituents of nuclei, are also considered as radiation and are relevant in radio-protection in nuclear facilities. Also streams of protons may be relevant for radioprotection in certain research facilities.

Like electromagnetic radiation, particles are relevant for risk assessment and management for the potential risks related to their capability to interact with matter and release energy.

The energy of particles depends on their mass and speed. Their capability to interact with matter depends also on their charge. In particular, Alpha particles, which are charged and are relatively massive, interact so strongly with the atoms of the matter across which they travel that they deposit all their energy in a very short path inside any type of matter.

Sources of Radiation There are both natural and artificial sources of radiation. We are literally immerged in a sea of electromagnetic radiation of different frequencies from natural sources: light and heat from the sun, infrared radiation from any "hot" body, radio-waves from space etc.

Electromagnetic radiation is generated by moving electric charges. A variety of devices generate electromagnetic waves, intentionally or as an unintended consequence of their functioning. Unintended emissions of electromagnetic radiation may create unwanted interference, disturbance or noise for communications or for the functioning of other devices. They may cause safety risk due to induced malfunctioning of devices or apparatus. Therefore, unintended emissions are subject to EU legislation on electromagnetic compatibility aimed at preventing or limiting dangerous interferences.

Radio-waves are extensively used for telecommunication. X-rays are produced by electron excitation, notably in devices built for medical diagnostics.

The most energetic electromagnetic radiation, gamma-rays, is generated in nuclear processes that are phenomena taking place inside atomic nuclei. Nuclei of heavy atoms, like Uranium 235, are subject to fission, when bombarded with neutrons, splitting into lighter more stable fragments with release of energy as gamma rays and high speed neutrons (and of alpha and beta particles from the fission products) (Fig. 4.7).

Similarly, nuclei of certain light atoms can combine under appropriate conditions in a process of fusion to form heavier nuclei with release of high energy radiation (Fig. 4.8).

Moreover, some elements which are naturally present on earth are unstable. They decay emitting gamma rays, alpha and/or beta particles. As fusion takes place in the

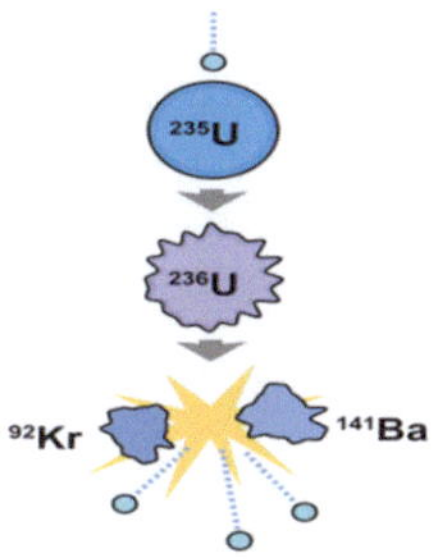

Fig. 4.7 The fission mechanism (Source: Wikimedia Commons (public domain))

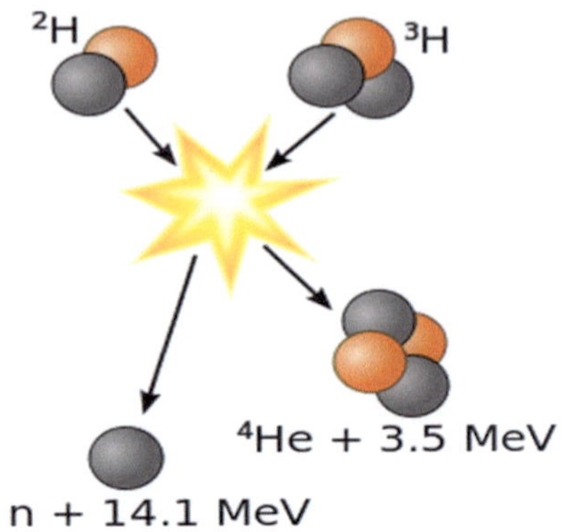

Fig. 4.8 The fusion mechanism (Source: Wikimedia Commons (public domain))

sun, the sun is a strong natural source of gamma rays, most of which are fortunately absorbed in the high atmosphere. Alpha and Beta particles are emitted from natural or artificial radioactive isotopes.

4.9.3 Interaction of Radiation with Matter and Related Potential Health Hazards

As said above, the interaction of electromagnetic radiation with matter depends on its frequency. In that respect, a fundamental distinction must be done between ionising and non-ionising radiation.

Below a certain threshold of energy, generally assumed to be 10 electron-volts (eV) radiation is not capable of extracting an electron from any type of atom. 1 eV is the energy acquired by an electron accelerated in a difference of potential of 1 V.

In practice, radiation below the "ultraviolet" frequency is considered to be non-ionising. Still, such lower-energy radiation has effects on matter, including biological effects on living matter, and even a potential for adverse health effects, depending on its frequency and intensity. The use of lower-frequency electromagnetic radiation is indeed motivated by its capability to produce useful effects like for example modulating signals or activating electric circuits.

Gamma Rays High energy gamma rays are very penetrating. They can easily pass through the human body and are only absorbed by thick concrete walls or lead screens. They are dangerous both for internal and external exposure.

X-Rays X-rays are less energetic and less penetrating than gamma-rays. Still, they can pass through the human body and are therefore extensively used for medical imaging.

Alpha Particles Alpha particles may be as energetic as gamma rays. However, since they are ions that interact strongly with the atoms of the matter they pass through, they have a very low penetration power and are arrested by a few centime-

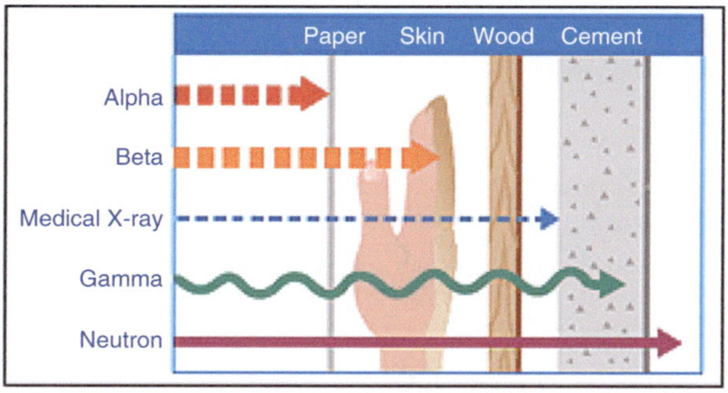

Fig. 4.9 Penetration power of different forms of radiation (Source: US NRC (public domain))

tres of air, by a sheet of paper or the outer dead layer of skin. Therefore, they are dangerous only when the source is inhaled or ingested because they may in such cases produce heavy localised damage.

Beta Rays Finally, Beta rays are more penetrating than Alpha ones, can penetrate the human body, but have a lower interaction potential and their deposition of energy is therefore more widely spaced and causes less harm in case of internal contamination by beta-emitting sources (Fig. 4.9).

Non-ionising Radiation On the current state of scientific knowledge, non-ionising radiation is considered of limited health relevance. Nevertheless, above certain intensity, very low frequency radiation can cause nerve stimulation, while at higher frequencies (radiofrequencies) it may give rise to tissue heating. Exposure limit-values are intended to prevent such proven effects. There is debate on possible health effects of low frequency electromagnetic fields. The matter is further discussed in the section devoted to risk management of non-ionising radiation.

Visible Light Radiation in the range of visible light is notoriously not penetrating and may cross only transparent or trans-lucid materials. Nevertheless, under certain conditions it may cause adverse effects on eyes, like in particular the development of cataracts or other eye pathologies.

Stochastic and Non-stochastic Health Effects; Dependence on the Dose In the case of ionising radiation of higher energy, a distinction must be made between stochastic and non-stochastic effects. The distinction is based on the level of radiation absorbed, the "dose".

The concept of *radiation dose* is different from and more complex than that of toxicological or pharmaceutical dose for chemicals or medicines. Three concepts intervene in the risk-relevant definition of radiation dose, summarised in Table 4.4:

- Absorbed dose: this is the energy deposited in a unit of volume of the exposed human tissue. The usual unit of measurement for the absorbed dose is the milligray (mGy).
- Equivalent dose: the equivalent dose quantifies the damaging potential of the specific type of radiation absorbed as different types of radiation have different properties. The unit of measurement is the milliSievert (mSv).

Table 4.4 Dose quantities

Absorbed dose
Energy deposited in a kilogram of a substance by the radiation
Equivalent dose
Absorbed dose weighted for harmful effect of different radiations (radiation weighting factor wr)
Effective dose
Equivalent dose weighted for susceptibility to harm of different tissues (tissue weighting factor wt)

- Effective dose: in addition to the energy deposited and the properties of the radiation considered, the effective dose takes into account the sensitivity of the exposed tissue or organ. It is again measured in mSv.

Different units are sometimes used in the United States for the same concepts: the rad, rem and roentgen.

Stochastic Effects Stochastic effects are those associated with chronic, long-term but low-level doses. The term stochastic implies that such effects can only be assessed in a probabilistic manner. Stochastic effects of high frequency radiation exposure are cancer or mutations. Mutations consist of changes in the DNA that may be teratogenic (inducing defects in the offspring) in case of exposure of the foetus, or genetic (changes in the genes which may be passed on from parent to child).

The *type and severity* of stochastic effects do not depend on the level of exposure. However, the *likelihood* of the effects increases with such level of exposure. There is uncertainty on the dose-effect relationship at low-dose levels, including the existence of a threshold. The issue is discussed in more detail below.

Non-stochastic Effects Non-stochastic effects are associated with high level of doses, concentrated over short periods of time (acute exposure). Even if different organisms may to a certain extent react differently to the same level of exposure, the effects can in general be assessed deterministically.

Non-stochastic effects include burns, radiation sickness and even death caused by heavy damage to tissue and organ due to very high energy release. Such high levels of exposure can only be expected for the public in case of major accidents, nuclear war or terrorist acts, while accidental exposure in the workplace may happen in nuclear industry or other activities involving radionuclides.

Lower Frequency Radiation Below the frequency of the ultraviolet radiation, electromagnetic waves do not carry sufficient energy to ionize matter and break chemical links. In particular, radiation in the ranges of radio frequency (RF) (100 kHz $< f \leq 300$ GHz), intermediate frequency (IF) (300 Hz $< f \leq 100$ kHz), extremely low frequency (ELF) ($0 < f \leq 300$ Hz), is non-ionizing. The damage mechanisms that are typical of ionizing radiation do not intervene at such lower frequencies.

Table 4.5 Meaning of prefixes used for expressing physical quantities

Femto	Pico	Nano	Micro	Milli		Kilo	Mega	Giga	Tera	Peta
10^{-15}	10^{-12}	10^{-9}	10^{-6}	10^{-3}	10^{0}	10^{3}	10^{6}	10^{9}	10^{12}	10^{15}

It is interesting to note that that the frequency of visible light is much higher that radio-frequencies. The human eye can see light in the wavelength range from about 390 to 700 nm that in terms of frequency corresponds to a band of 430–790 THz. 1 THz is 1000 GHz as shown in Table 4.5, which clarifies the meaning of prefixes widely used in physics for practical reasons:

Radio frequency (RF) fields are particularly important for their many applications in today's communications. Sources include in particular mobile phones, cordless phones, local wireless networks and radio transmission towers. RF fields are also used for medical scanners, radar systems and microwave ovens.

Extremely low frequency (ELF) fields, below 300 Hz are generated by the alternating currents in power lines, wiring and appliances. Fields generated in the proximity of transmission lines are particularly strong.

Exposure to non-ionising radiation is subject to international and EU guidelines and is regulated at national level. The EU has adopted a non-binding recommendation which is however the basis for binding product standards. As explained before, the exposure limit values are based on the acute, observed effects and include substantial safety factors. For instance, for mobile phones the EU has established a limit for the human head (which is the most exposed part of the body) at a Specific Absorption Rate (SAR) of 2 W per kilogram of tissue. Since mobile phones are tested assuming worst-case conditions and only a power hundreds or thousands of times lower than the maximum power conventionally assumed for the test is actually transmitted during the use of a phone, the exposure under real conditions is generally much lower than the applicable limit.

Other wireless devices, cordless phones etc. cause exposures which are in general lower than that caused by the use of mobile phones.

Also the exposure to extremely low frequency fields is below the set limits and even in the proximity of high voltage power lines, the sources of strongest ELF, exposure remain below the applicable safety limits.

In spite of the existing limit values and standards for the exposure to low frequency radiation, and the findings of studies and assessments that so far have not detected negative health effects, concerns are often expressed on potential health risks of mobile phones, masts and transmission lines. Things are complicated by some epidemiological, weak evidence of an increased incidence of leukaemia in children exposed to strong ELF magnetic fields from power lines. However, a causal relationship between exposure and leukaemia is not confirmed by other lines of evidence (animal or in vitro tests). For other effects, *fatigue*, headache, and concentration difficulties, generally self-reported, no relation with extremely low frequency fields has been demonstrated either.

In the absence of a robust basis of actually observed systemic effects for identifying and characterizing health hazards, beyond those hazards which are related to immediate, acute effects, risk assessment has no means to provide information on "safe" exposure levels addressing effects that remain hypothetical. Prospective cohort studies have been launched in order to provide better evidence. As far as research is concerned, the complex cell-functioning mechanisms on which science is progressively shedding light are mentioned as a possible direction for exploring ways in which non-ionizing radiation could interfere with the functioning of cells.

4.9.4 Radiation Risk Assessment

Radiation risk assessment goes through various steps conceptually similar to those applicable to the health risk assessment of chemicals.

Identification and Characterisation of the Radiation Health Hazard The starting step is the identification and characterisation of the radiation health hazard. This involves the identification of the type of radiation involved and the determination of the type of health damage that it can cause at the different exposure levels.

We have already highlighted the different impacts due to high or lower level exposure to the same type of radiation. A distinction is made between short term, acute effects of high level exposure, that can be assessed in a deterministic way, and longer term, stochastic effects related to lower level exposure that are assessed probabilistically. To what already said on stochastic effects, we should add that they are generally considered to show no threshold, with a linear relationship between dose and effect. Nevertheless, there is no consensus on the existence of effects at very low level of exposure.

The Assessment of Exposure A second step is the assessment of exposure. It may be very difficult in many cases, for instance when modelling the exposure of the population from a major nuclear accident, to determine the radiation doses received by the various individuals and members of the different population groups. Things may be complicated further when, in addition to external exposure (exposure from radiation sources outside the body), assessors have to account also for internal exposure from ingested or inhaled radioactive contamination.

Computer software programs have been developed for calculating the doses from the intake of various radionuclides. In such cases, knowledge of the chemical characteristics of the contaminants is also important as it determines which are the organs and tissues directly exposed.

The Characterisation of Radiation Risks Finally, assessors proceed to the characterisation of radiation risks combining the data on exposure with the information characterising the radiation hazard. Long term radiation risks are normally expressed in terms of probabilities and expected effects, for instance the expected increase in the number of additional cancer cases per 10,000 or 1,000,000 people exposed. Often a range is indicated instead of a single figure, with the relevant distribution of probability. Cancer risk from low-level radiation exposure may be indicated in absolute terms (for instance a total lifetime incidence of cancer of 1 in a 1000 people) or as relative risk expressed as a factor multiplying the baseline risk in a non-exposed population (like a relative risk of 1.5 multiplying the background risk of, for instance, 2 cancers per 1000 people: $1.5 \times 2 = 3$ cancers per 1000 people).

Although the conceptual approach of radiation risk assessment is straightforward, its practical implementation requires the application of complex modelling techniques and significant computer power, depending on the accident and exposure scenario. In most cases, radiation risk assessment is affected by large uncertainties and can only lead to a probabilistic expression of risk.

The assessment paradigm exposed in this section needs to be adapted to the actual circumstances and to the objectives of specific assessments. The assessment may be prospective, trying to model what could be the consequences of a hypothetical future accident leading to the exposure of a particular group or the wider population, or retrospective, aiming to estimate the long term impact of an actual exposure situation. It may apply to the external exposure from contaminated sites or to the effects of intake of radionuclides or both.

Steps of Radiation Risk Assessment Conceptually, the steps of the assessment are very similar across the different situations. In the particular case of workplace exposure to radiation risk, the steps of an assessment may be reformulated in the following way:

1. Identification of the sources and circumstances of possible exposure and of the types of nuclides and radiation involved
2. Characterisation of all possible radiation hazards.
3. Identification of all workers and other people who may be exposed to the radiation hazards identified
4. Determination of the possible exposure levels of workers and other people under the various possible circumstances and account taken of the safety procedures and protective measures in place
5. Characterisation of risks in terms of the likelihood of the hazard occurring and the severity of health damage that may result.

4.10 Safety Assessment of Nuclear Power Plants

4.10.1 Introduction

Nuclear power is a subject of debate and controversy in the EU and worldwide. A few EU Member States have invested heavily in the nuclear power, which represents a significant share of their electricity production. Some Member States have banned it altogether, while a few have decided to stop their nuclear programmes and progressively close their nuclear plants. The nuclear issue re-emerges periodically in the policy debate on climate change and the security of energy supplies. Nuclear power plants produce electricity without CO2 emissions (although a comprehensive balance should take into account all the energy input going into the building, operation and decommissioning of a nuclear plant, accounting for the possible associated greenhouse gas emissions). Nuclear power is therefore often referred to as possible part of a "de-carbonised" energy policy.

Establishing the balance between the pros and cons of nuclear is a very complex exercise, involving difficult technical analysis and value judgements. Possible risks, related to the entire nuclear industry, including fuel production, power plant operation, transport of nuclear materials and nuclear waste disposal, are at the core of the debate. Other important aspects are nuclear weapons proliferation, energy security and the energy system models with all their economic and societal implications. The vulnerability of nuclear facilities to terrorist attack is also a reason for concern.

Our objective here is to introduce the broad lines of the risk assessment methodology for the particular segment of the nuclear cycle represented by the nuclear power plants. The specific interest of risk assessment of nuclear power plants for this chapter devoted to risk assessment is that a technical approach has been developed specifically for these risks, which addresses the probabilities of events leading to adverse outcomes in a systematic and quantitative way. This approach is typical of engineering risk assessment.

Since the political debate on nuclear power is inevitably part of the deliberations on the EU energy systems, notably in relation to the de-carbonisation of the economy, it is important to provide here some background on the safety of nuclear power plants and on the probabilistic risk assessment methodology (PRA). PRA is relevant for a large number of risks posed by technical systems in many engineering areas, like for example IT and energy systems or aviation safety. The presentation given above on radiation risks is obviously very relevant for understanding the risks of nuclear power plants, which may end up in radiation health risks for workers and the public, in physical damage to equipment, in loss of infrastructures and property as well as radioactive contamination of environmental media across wide geographical areas.

Engineering risk assessment aims at identifying as completely as possible the initiating events that, alone or in conjunction with other events, can lead to an accident happening and to a hazard materialising, often through a cascade of subsequent events. Since the technical systems to which engineering risk assessment is applied, like in particular nuclear plants, are usually extremely complex, sophisticated techniques have been developed in order to systematically calculate the overall probability of accidents from the probabilities of the individual events in a chain.

In spite of a long experience with probabilistic risk assessment (PRA) and much research and development work, some intrinsic limitations affect this approach. It is very challenging to identify all the possible initiating events and their ramifications, to model the possible cascades of events and the complex interactions between components of a system. The knowledge basis for assigning probabilities to the relevant events is sometimes quite weak. The human factor is important but difficult to model and to quantify. In particular, accidents like the one of Three Miles Island (described later in this section) show the importance and unpredictability of human mistakes in the human-machine interaction. Anticipating natural events has also proven problematic, as shown by the Fukushima accident. Finally, the threat of terrorist acts represents a new challenge that was not addressed in the traditional PRA studies.

Technical assessment can provide decision makers with a robust framework for rational reasoning on risks in this important area and for identifying measures aimed to reduce those risks, but it is illusory to expect from "science" a complete assurance of safety or a single-figure measure of risks posed by complex systems.

4.10.2 The Scientific Bases of Nuclear Power

Current nuclear power plants technology is based on controlled fission of Uranium nuclei. Matter is composed of atoms that in turn are made of different sub-atomic particles: nuclei composed of protons and neutrons and, around the nucleus, electrons. Protons, neutrons and electrons have already been introduced in the part of this Chapter dealing with radiation. Let's just say that the complete atom is electrically neutral since the total negative charge carried by the electrons is compensated by the total positive charge of the protons. Neutrons and protons are kept together in the nucleus by nuclear forces that are much stronger than the electrostatic repulsive force between the positive protons that would tend to disintegrate the nucleus. However, when heavier atoms are considered, with a great number of protons in the nucleus, an even greater number of neutrons are required in order to "bind" the nucleus together. Above a certain size, the nucleus becomes unstable.

The Fission Mechanism The atom of Uranium is a very big one with 92 protons and 92 electrons. The number of neutrons may vary since Uranium has several isotopes, atoms with the same number of electrons and protons and different number of neutrons. The most abundant isotope in nature is U238 with its 92 protons and 146 neutrons. A small percentage of the Uranium in nature has a slight lower number of neutrons: U235, which is present as a 0.72 % fraction of total Uranium, has 143 neutrons in its nucleus (Fig. 4.10).

The important property of U235 is that it can capture under certain conditions a neutron passing near to its nucleus, and the capture may be followed by the fission of the nucleus with the release of energy (Fig. 4.11).

The probability of the fission happening increases when the incoming neutrons are relatively slow. The neutrons emitted by the fission itself (2.5 on average from each fission) are very fast. Therefore, they must be slowed down (moderated in the nuclear engineering jargon) in order to be able to cause fission in other U235 nuclei at a high rate. In that way, provided the density of U235 atoms is sufficient, a self-sustaining chain reaction can take place (Fig. 4.12).

Fig. 4.10 Constituent particles of the atom of uranium and its nucleus

NUCLEOUS		ELECTRONS
Protons	Neutrons	
92	143 or 146	92

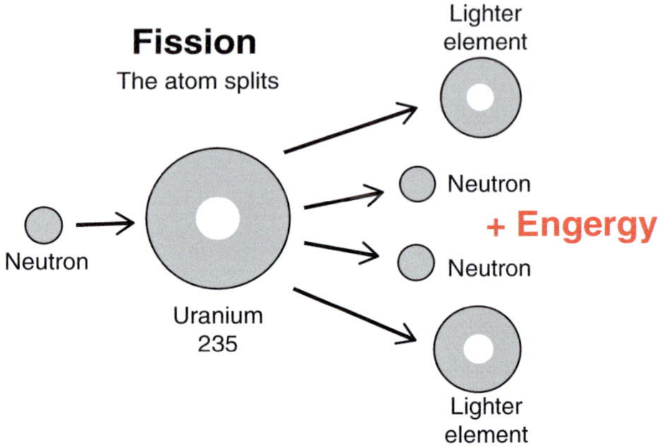

Fig. 4.11 The process of nuclear fission (Source: US National Energy Education Development Project (public domain))

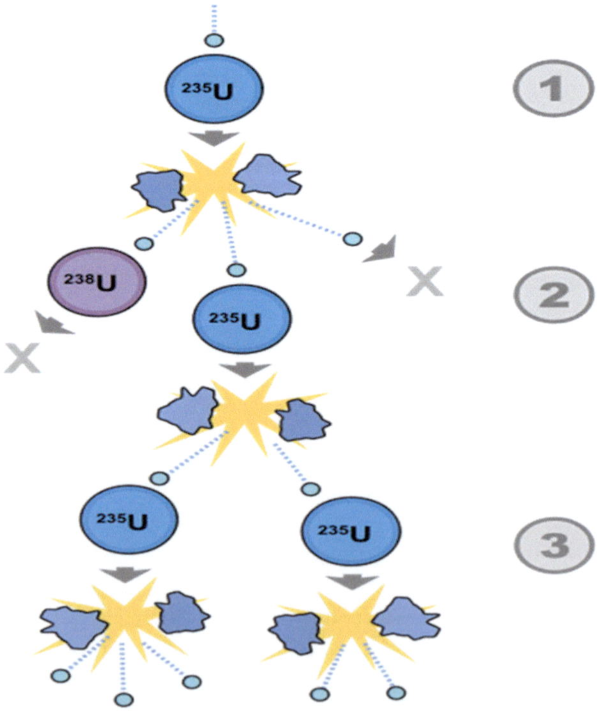

Fig. 4.12 Self-sustaining chain reaction (Source: Wikimedia Commons (public domain))

Water is used to slow down fission neutrons. When neutrons collide with the Hydrogen nuclei of water, they lose part of their kinetic energy. Since water also absorbs neutrons, the nuclear fuel must be enriched in U235 in order to sustain a controlled chain reaction.

Energy from the Fission The fission of one U235 nucleus releases about 200 MeV (two hundred million eV) and about 85 % of this energy is in form of kinetic energy of the fission fragments. The kinetic energy of fission fragments is released into the matrix of the fuel and converted into heat. The amount of energy released in nuclear fission reaction is huge compared to the energy from combustion of fossil fuels: the fission of the nuclei in 1 g of Uranium produces the same energy as the combustion of approximately 3 tonnes of coal.

4.10.3 Functioning of Nuclear Power Plants

Nuclear power plants produce electricity in the traditional way, with a steam turbine and an electricity generator, exactly as in any fossil fuelled power plant. The difference resides in the way the steam is produced. In a nuclear power plant it is the heat from nuclear fission, instead of that from coal, oil or gas combustion, which is used to produce steam. There are several types of nuclear power plants. There is no need to present here the details of the various plant designs, since the relevant risk concepts can be adequately illustrated with reference to the most common types of plant present in the EU.

The Pressurised Water Nuclear Power Plant The most important design for the EU situation, the Pressurised Water Reactor (PWR), is represented in Fig. 4.13.

The nuclear section of the plant is contained in a building, where are situated both the primary containment vessel, a steel vessel with the nuclear fuel inside, and the steam generator, where the water coming from the vessel arrives at high temperature and pressure into a heat exchanger that generates the steam going to the turbine via a secondary circuit.

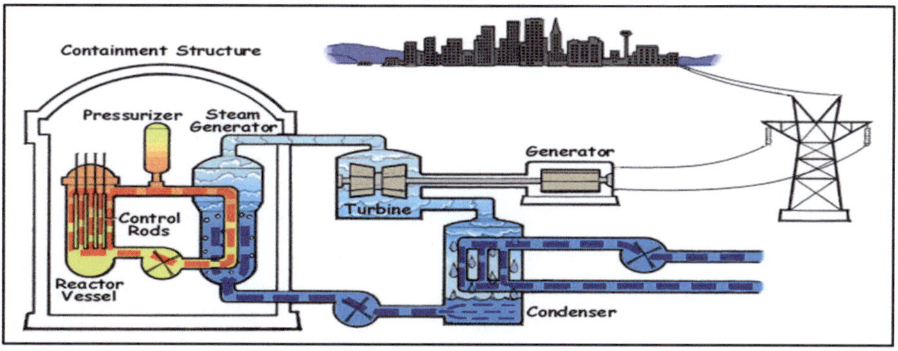

Fig. 4.13 The pressurised water nuclear power plant (Source: U.S.NRC "The Pressurized Water Reactor" (public domain))

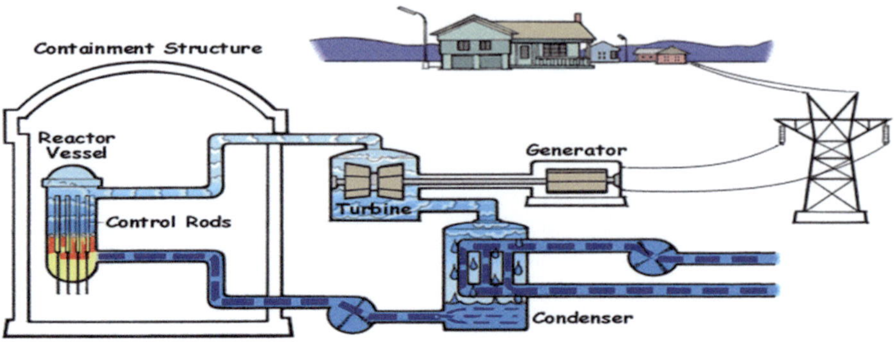

Fig. 4.14 The boiling water nuclear power plant (Source: U.S.NRC "The Boiling Water Reactor" (public domain))

The Boiling Water Nuclear Power Plant An older design, the Boiling Water Reactor (BWR), still present in the EU, has a single circuit, with the steam directly coming from the vessel as shown in Fig. 4.14. Such different design has important safety implications, notably because the steam is produced with water in direct contact with the fuel (which is however contained in cladded rods). Because the water in contact with the nuclear fuel becomes contaminated by radionuclides, contamination is transferred to the turbine. The turbine must be shielded and protective measures must be applied during maintenance.

Containment of Nuclear Material The nuclear reaction takes place in the reactor vessel, a massive steel container where the fuel rods are situated. The water in the vessel has the double function to slow down the fission neutrons and to cool the fuel, removing the heat produced by fissions and transporting it to the steam generators (Fig. 4.15).

The nuclear fuel is normally constituted of pellets of Uranium Oxide (UO2), contained in assemblies of rods with metallic (Zirconium) cladding (Fig. 4.16).

4.10.4 Safety Aspects of Nuclear Power Plants

The most important safety objectives for a nuclear power plant are shielding the radiation produced by the fission process and the fission products and containing the radioactive material produced in the nuclear fuel under all circumstances.

Protection against radiation, mostly neutrons and gamma rays (since the fission products themselves lose their energy within the fuel itself) is ensured by the water in the vessel and by massive concrete containment and protection structures.

Barriers to Prevent Release of Radioactive Material The nuclear fuel is physically contained by the steel vessel that is designed to stand very high temperature and pressure. All the nuclear section of the plant is isolated within a reactor building kept at a pressure lower that the outside atmospheric pressure.

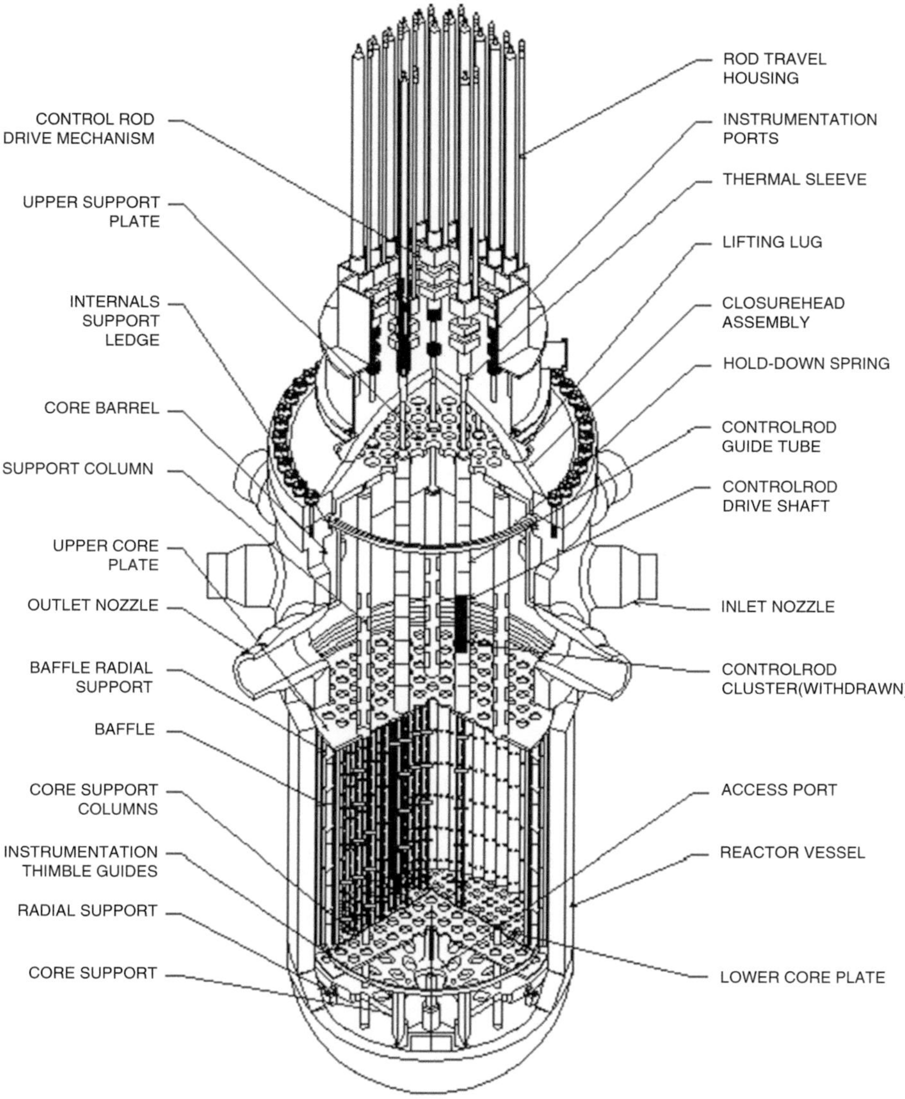

Fig. 4.15 The vessel and the core of a nuclear power plant (Source: US Energy Information Agency (EIA) (public domain))

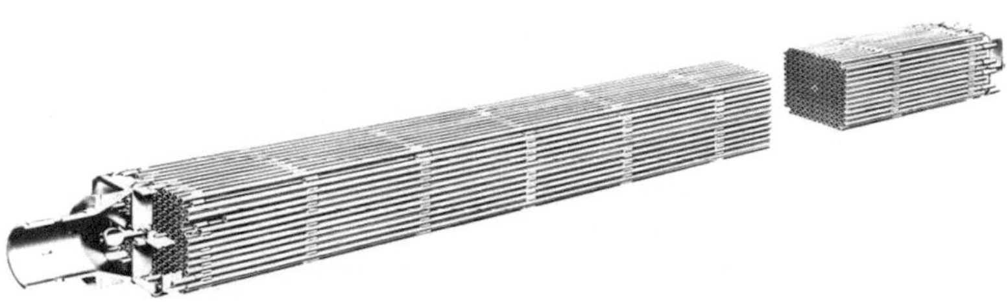

Fig. 4.16 The fuel element of a nuclear power plant

Several barriers contribute to prevent release of radioactive material from a nuclear power plant. The fuel itself is self-contained since the radioactive fission products, which are produced inside the UO2 pellets, under normal operating conditions remain inside the solid ceramic matrix. The pellets are contained in sealed fuel rods of zirconium alloy. The rods, clustered in assemblies, are contained in the steel pressure vessel. The vessel, with steel walls up to 30 cm thick, is placed in a containment building with very thick walls of concrete (normally more than 1 m).

Hazards and Safety Issues in a Nuclear Power Plant A most important safety objective in a nuclear power plant is to maintain the cooling of the core under all circumstances. Even after the nuclear reaction is stopped (by the control rods that can be inserted into the core and are made of material absorbing the neutrons, and/or by a loss of coolant-water, in the absence of which neutrons are not slowed down and the reaction cannot be sustained), an important amount of heat is produced in the fuel by the decay of fission products, which goes on for a long time. In the absence of cooling, the core may be damaged by the excess heat and can even melt down.

In a reactor of the types seen above, there no risk of nuclear explosion, because the concentration and configuration of fissile material, even after a possible core melt-down, is not sufficient to reach the mass and conditions similar to those of a nuclear bomb. Nevertheless, chemical explosions may take place under accident conditions, because of the possible accumulation of Hydrogen, which is produced at high temperature by reaction between the Zirconium in the fuel cladding and the coolant water.

The critical safety issues in a reactor, relevant for health and environmental protection, are therefore the possible damage and melting down of the core, caused by insufficient cooling, and the loss of nuclear material into the environment following loss of containment, which may be caused in particular by chemical explosions. For workers, risks are related to possible acute irradiation or contamination accidents, under normal operating conditions of the plant itself or under accident or emergency conditions.

There is a variety of accident-initiating factors in a nuclear plant, including for example inadequate design, human error, natural events like earthquakes, equipment malfunctioning, structural breakings, terrorist acts etc.

It must be noted that, in addition to the installation necessary to transform nuclear energy into electricity, are also normally located in the site of a power plant storages or fresh fuel, of nuclear waste, and a pond where spent fuel is kept under water. These facilities involve their own specific safety aspects. There are many other indirect safety aspects linked to the operation of a power plant, like the treatment and disposal of its nuclear wastes, the transport of nuclear material, the manufacturing of fuel, non-proliferation issues etc. For our objective, which is to introduce the probabilistic risk assessment methodology, it is not necessary to treat all of them here. We focus on the safety aspects of the sole power production section.

4.10.5 Probabilistic Risk Assessment (PRA) of Nuclear Power Plants

The Probabilistic Approach to Risk Assessment Safety assessment and management of nuclear power plants aim to ensure, based on comprehensive assessment of risks, that the operation of such plants do not involve unacceptable public health and safety risks. A nuclear plant is an extremely complex technical system. A very structured and systematic approach is needed in order to identify its hazards, calculate risks and provide a basis for establishing the appropriate technical and operating requirements. To that aim, a probabilistic risk assessment (PRA) method is applied in order to systematically analyse the functioning of such complex system and its dependency on its components. Under the PRA approach, risk involves consideration of hazards and their frequency and of the severity of the potential consequences when hazards materialize.

Once a particular hazard is identified, the objectives of the safety analysis are to determine the possible frequency of a particular hazardous event and its potential adverse consequences,

Therefore, in order to reduce risks, measures may be taken either to lower the likelihood of the dangerous events or to mitigate the gravity of their consequences, or by acting on both aspects.

Steps of PRA A probabilistic risk assessment includes the following steps:

(a) *Hazard identification*: the major hazard in a nuclear plant is the damage of the core, in particular its melting down.
(b) *Identification of the initiating events*, technical, natural or human, that can cause the hazard to materialize, in particular those that can lead to the damage of the core. Examples are failure of pumps or breakage of a coolant pipe.
(c) *Determination of the probability associated to the initiating events* which have been identified. This is based on technical analysis and consideration of experience with the frequency of failure of equipment. It is particularly difficult to estimate the likelihood of initiating events linked to human or natural causes.
(d) *Determination and analysis of the cascades of possible secondary events*, malfunctioning and failures of components, and of their interaction and combination, following the initiating events that may lead to negative outcomes.
(e) *Calculation of the overall probability for the negative outcome* is the final step. That is often the result of complex calculations combining the probabilities of initiating events and the cumulative probability of the cascades of sub-events and failure leading to the negative outcome.

In order to model the complex systems of a nuclear power plant and the events and faults that can lead to certain outcomes, event trees and fault trees are used.

Event and Fault Trees A very simple example of an event tree is shown in the scheme of Fig. 4.17 that also illustrates how the probabilities are combined:

The combination of probabilities determined by this modelling method provides the overall probability of the outcome.

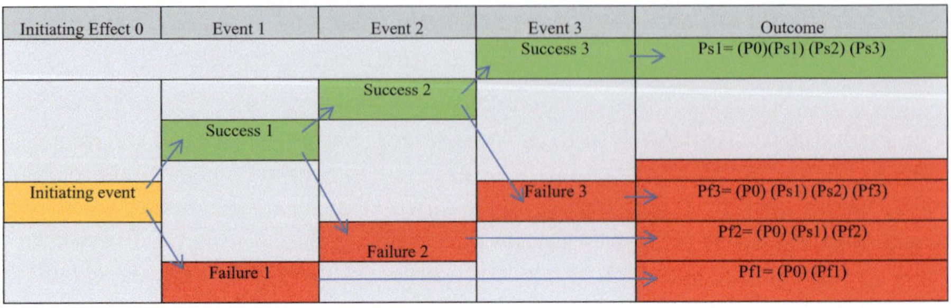

Fig. 4.17 Event tree

As said before, human causes are particularly difficult to analyse and quantify. The technique used, Human Reliability Analysis (HRA) tries to assess the probability of human errors of the operators, taking into account operating procedures, level of training, and other relevant conditions.

PRA is a probabilistic approach that allows dealing systematically with uncertainties, variability and incomplete knowledge. The so-called Monte Carlo methods are applied to probability distributions for the factors subject to uncertainty and variability. This technique allows computing distributions of probability for the possible outcomes by combining in a repeated way random chosen values from the distributions of probability of the events and sub-events in the relevant cascades.

Probabilistic risk assessment of nuclear plants requires a lot of computer power in order to establish, refine and run the models needed to analyse the many possible events and their consequences. In order to simplify the approach, the assessment considers separately the internal events, which are related to the internal systems of the plant, and the external ones, which cover natural events like earthquakes, flooding, storms, tides, tsunamis or hurricanes.

Levels of PRA Conventionally, PRAs is organized in several levels:

Level 1
The objective is to provide an estimate of the frequency of core damage, taking into account all the events affecting the systems that protect the reactor and the state of operation of the plant. The uncertainties related to this stage of the assessment can be kept at a low level by careful modelling of the internal systems of the reactor
Level 2
The objective is to estimates the magnitude of the radioactivity that can be released following the core damage, as well as the time-evolution of the accident. There are significant uncertainties in this stage since the efficacy of the containment as well as the loss of coolant are subject to large variability
Level 3
The objective of level 3 is to assess the consequences, for health, property etc., of the core damage and of the resulting release of radioactivity from the containment. Health damages and economic losses are evaluated. Here, results are affected by the great variability of the relevant factors including the meteorological conditions, in particular wind speed and direction, during the period of release

As said above, the result of PRA takes the form of a range of outcomes with the related distribution of probabilities. Refinements to the modelling, better understanding of the underlining physical processes, more complete identification of initiating effects etc. can reduce the level of uncertainty. Nevertheless, due to the limitations of the assessment method, in order to ensure the level of safety required with adequate confidence, large safety margins are used.

4.10.6 Lessons from Nuclear Accidents

The Chernobyl Accident The 1986 disaster at the Chernobyl nuclear power plant has shown the importance of the safety standards that prevail in most countries, but were not implemented in Ukraine. The RBMK type of reactor that was involved presented major design deficiencies. In the RBMK design the loss of coolant could increase the power output instead of arresting the fission reaction. Moreover, operating procedures were violated. Overall, there was little attention to safety requirements. Due to the particular conditions of the Chernobyl disaster, it is not considered particularly relevant for analysing the safety of most commercial plants.

The Fukushima Accident The Fukushima accident in March 2011 provides more relevant lessons. The nuclear reactors at Fukushima were shut down automatically following the March earthquake. After the shut-down, the back-up generators ensured the electricity needed to power the pumps allowing the water circulation for residual heat removal from the core of the reactors. Unfortunately, when the tsunami struck the plant, the back-up diesel generators for three units were swamped. The emergency core cooling systems then failed. The height of the tsunami hitting the plant was 14 m. Protection against tsunami had been designed for a maximum height of 5.7 m for Daiichi and 5.2 m for Dini, the two units that suffered flooding of pump-rooms. As a result the cooling of the reactor by the heat removal circuit was lost. In a sequence of events, additional problems came from the spent fuel ponds that lost water, and the explosions due to Hydrogen produced at high temperature by a reaction between the Zirconium claddings and water.

A number of weaknesses have appeared from the Fukushima accident: inadequate design of the protection against tsunami, a wrong siting of the plant, inadequate design of the back-up power, inadequate design of venting of the containment, inappropriate positioning of the spent fuel pond etc. Those weaknesses show the difficulty of determining the nature and size of the events that can lead to damage as well as of establishing the appropriate, adequate design specifications.

The Three Mile Island Accident Another, older example of important weaknesses in the design and operation of a nuclear reactor is provided by the Three Mile Island Unit 2 (TMI-2) accident. The plant was situated near Middletown, Pa., US. Its core partially melted down in March 1979. The resulting radioactive releases were small and without detectable health effects on plant workers or the public. The economic losses and the impact of the public opinion were instead huge. The accident was due to a complex sequence of technical failures and human operator misunderstandings and mistakes. Inadequacies emerged on various aspects: emergency response planning, reactor operator training, human factors engineering, radiation protection.

Lessons from accidents and near-misses are continuously incorporated into assessment methods and safety standards. Overall, it is considered that the safety assessment of nuclear plants and the safety design and operation standards adopted for commercial nuclear plants have guaranteed a high level of safety. Nevertheless, the few serious accidents that have taken place show that current nuclear technology and the safety assessment methods are not without weaknesses.

Summary

1. Risk assessment includes some basic steps that are similar across the different risk sectors: the identification and characterisation of hazards, the assessment of the exposure to the stress factors, the final quantitative or qualitative characterisation of the risk. Risk assessment focusses on the technical dimension of risk. It must therefore be complemented by a broader appraisal of the wider societal aspects. Technical risk assessment applies scientific tools and is based on a structured, systematic approach. However, there are limits to a scientific approach to risk and transparency on certain subjective aspects of risk assessment is key for the credibility of the process.
2. Various problems are common to technical risk assessment performed to inform the policy and regulatory process. The health risk assessment of chemicals offers a useful paradigm in that respect. It involves the use of several scientific disciplines and its process and results are in general expressed in technical terms that many decision makers without specific background do not easily understand. For the same reason, it is also difficult for many to grasp the limits of science in assessing risks, the reasons for divergent opinions and the influence of uncertainty on risk evaluation.
3. Risk assessment of chemicals is based on well-established methods. Nevertheless new knowledge and tools, notably the so-called "omics" and the available data treatment capacity open new perspectives but also pose new problems for risk assessment, notably the interpretation of the health relevance of effects on the cell and molecular level.
4. Chemical hazards are studied by various methods, on the level of living organisms with animal tests, on tissue and cells (in vitro test), by studying the relevant aspects of their chemical structures using informatics tools and by epidemiological studies, observing their effects on exposed groups. Animal models are currently the main source of data for health risk assessment as they allow observation on an entire organism.
5. Hazard categories are defined in the EU legislation, in line with an international classification system. Hazard are characterised through the determination of the relation between the dose and the effects in animal tests. It is assumed that certain chemicals show effects for which there is no lower threshold whereas in

other cases it is considered that below a certain dose-threshold the effects are negligible.

6. In order to establish the risk for human health posed by chemicals, the observations on animals are transposed to humans by applying safety/uncertainty factors, most often based on conventional assumptions. The resulting risk for human health associated with exposure to a given chemical is determined by coupling the information or assumption on exposure with the data derived from animal tests. Various technical expressions of risks are used. For threshold effects, results of risk assessment are in general not quantified and expressed in qualitative terms only (negligible, serious etc.). For non-threshold effects, notably genotoxic carcinogenicity, the expected incidence of additional cancers associated with a given level of population exposure may be estimated or, alternatively, may be indicated the exposure level below which the additional incidence is lower than an "acceptable" level.
7. Several more specific issues for the risk assessment of chemicals are still open. Often only a partial view of the potential health impact of a given chemical is provided considering the exposure from a single source and exposure route like for example food ingestion. There is no agreed and consistent approach for the assessment of mixtures of chemicals. There is no final conclusion on the existence of a threshold of effect or certain carcinogenic substances. Various, non-harmonised approaches are used for expressing the impact of uncertainty on the estimated risk. A certain degree of subjectivity is present in the way in which evidence and data are selected and "weighed", which may lead to divergent opinions.
8. The assessment of microbiological risks follows an approach similar to that developed for chemicals. However, microbiological risks are characterised by the ability of microorganisms to reproduce themselves, be subject to mutation and transfer of genes between different strands.
9. In the case of ecological risk assessment, specific parameters are introduced in order to study the effects of chemicals and other stressors on living organisms. Particular problems are posed by the complex relation between emission sources and the resulting exposure of the environmental targets, that must take into account dispersion, deposition, transportation, re-concentration, bioaccumulation across the various environmental media.
10. Risk assessment of food and feed derived from genetically modified plants follows the same approach applied to the health assessment of chemicals. Nevertheless, a specific concept of "comparator" is introduced, that is a crop for which there is a history of safe consumption, and the assessment approach is comparative and aims to determine whether the modified plant presents any difference with respect to the comparator, beyond the natural variability of the latter. The study focuses on the intended and non-intended effects of the modification, the composition changes, the newly-expressed proteins. It addresses the toxicity characteristics of the modified plant and its products, its possible allergenicity and its nutritional properties. The objective of risk characterisation is to determine whether in any of such respects, the modified plant is as safe, nutritious etc. as the "comparator".

11. In the case of radiation, the biological hazard depends on the type of the corpuscular or electromagnetic radiation considered and its energy. The energy transported by electromagnetic energy depends on its frequency. Although gamma rays have exactly the same nature of radio waves and visible light, their highly penetrating and hazardous properties are due to their much higher frequency. Specific definitions apply to the concept of dose making a distinction between the quantity of energy deposited in the exposed tissue (Absorbed dose), its damaging potential due to the characteristics of the specific type of radiation (Equivalent dose), the sensitivity of the exposed tissue (Effective dose). Radiation risk assessment follows an approach similar in its broad lines to that applied to chemical risk, including the characterisation of the hazard, based on consideration of the type and energy of the radiation, the assessment of the exposure and sensitiveness of the target and the resulting dose, leading to the final estimate of the resulting risk. Specific calculation methods are applied, but the ex-ante assessment of potential radiation risk from short and long term, direct and indirect, external and internal exposure to radiation following accidental released of radionuclides may be very complex and uncertain.
12. The safety assessment of nuclear power plants provides an example of engineering risk assessment based on a probabilistic approach. Nuclear plants integrate a number of active and passive safety features, including in particular multiple containment barriers to prevent release of radioactive material, redundant safety system and key technical components. The worst accident is represented by a loss of refrigeration leading to the melting of the nuclear fuel and the release of radioactivity to the environment. The approach to safety assessment is based on the identification of the sequences of events and faults that can lead to such an accident, the assignment of probabilities to such events and faults and the calculation of the cumulative probability of the resulting outcomes. The well-known accidents of Three Mile Island, Chernobyl and Fukushima have however shown the difficulty to anticipate all the possible sequence of events and, particularly in the case of Three Mile Island, the importance of the human factor.

References

Bergman, A., et al. (2013). The impact of endocrine disruption: A consensus statement on the state of the science. *Environmental Health Perspectives, 121*(4), A 104–105.

BfR-German Federal Institute for Risk Assessment. (2009). *Establishment of assessment and decision criteria in human health risk assessment for substances with endocrine disrupting properties under the EU plant protection product regulation*. Report of a Workshop hosted at the German Federal Institute for Risk Assessment in Berlin, Germany, from November 11–13, 2009.

Bradford Hill, A. (1965). The environment and disease: Association or causation? *Proceedings of the United Kingdom Royal Society of Medicine, 58*, 295–300.

Dietrich, R., et al. (2013, September 5). Scientifically unfounded precaution drives European Commission's recommendations on EDC regulation, while defying common sense, well-established science and risk assessment principles. *Chemico-Biological Interactions, 205*(1), A 1–5.
ECHA (European Chemical Agency). (2010). *Practical guide 2: How to report weigh of evidence*.
ECHA (European Chemical Agency). (2012). *Guidance on information requirements and chemical safety assessment-Chapter R.19: Uncertainty analysis*, 2012-1.
ECHA (European Chemical Agency). (2012). *Weigh of Evidence*, 2012-2.
EFSA (European Food Safety Authority). (2009). Use of the benchmark dose approach in risk assessment: Guidance of the Scientific Committee. *The EFSA Journal, 1150*, 1–72.
EFSA (European Food Safety Authority). (2010). Scientific report of the endocrine active substances task force. *EFSA Journal, 8*(11), 1932.
EFSA (European Food Safety Authority). (2013). Scientific opinion on the relevance of dissimilar mode of action and its appropriate application for cumulative risk assessment of pesticides residues in food. *EFSA Journal, 11*(12), 3472.
EFSA (European Food Safety Authority). (2013-1). Scientific committee; scientific opinion on the hazard assessment of endocrine disruptors: Scientific criteria for identification of endocrine disruptors and appropriateness of existing test methods for assessing effects mediated by these substances on human health and the environment. *EFSA Journal, 11*(3), 3132.
European Commission. (2012). *Communication on the combination effects of chemicals-chemical mixtures*. COM (252) 2012.
European Union. (2001). *Directive 2001/18/EC of the European Parliament and the Council on the deliberate release into the environment of genetically modified organisms and repealing Council Directive 90/220/EEC*.
Klimisch, H. J., Andreae, M., & Tillmann, U. (1997). A systematic approach for evaluating the quality of experimental toxicological and ecotoxicological data. *Regulatory Toxicology and Pharmacology, 25*, 1–5.
SCCS (2007). *Opinion on phthalates in cosmetic products*.
SCENIHR (Scientific Committee on Emerging and Newly Identified Health Risks). (2012). *Memorandum of the use of scientific literature for human health risk assessment purposes. Weighing of evidence and expression of uncertainties*.
SCENIHR (Scientific Committee on Emerging and Newly Identified Health Risks). (2014). *Opinion on the safety of medical devices containing DEHP plasticised PVC or other plasticisers on neonates and other groups potentially at risk*.
SCHER, SCCS, SCENIHR. (2012). *Opinion on the toxicity of chemical mixtures*.
UK Central Research Laboratory. (2007). *Comparative review of terminology, study made for the European Commission*.
US EPA (Environmental Protection Agency). (1998). *Guidelines for ecological risk assessment*.
US EPA (Environmental Protection Agency). (2003). *Mechanism and mode of dioxin action*.
US NRC (National Research Council). (1994). *Science and judgement in risk assessment*. Washington, DC: National Academies Press.
WHO/IPCS (World Health Organisation/International Programme on Chemical Safety). (2002). *Global assessment of state-of-the-science on endocrine disruptors*, WHO/IPCS/02.2.

Chapter 5
Risk Management

Abstract Although risk assessment and risk management are two distinct functions of risk analysis, close dialogue and collaboration between assessors and managers are essential. Since risk assessment is not pure science and necessarily involves value-laden assumptions and criteria, risk managers and risk assessors should jointly establish a risk assessment policy. Risk management is the process of identifying and assessing policy options in view of the results of risk assessment and in order to support decision making. The EU risk management process involves several layers across a complex institutional structure and is currently framed within a set of "Better Regulation" guidelines and practices introduced by the European Commission and shared with all the EU Institutions via an inter-institutional agreement. Some basic principles apply: subsidiarity, proportionality and precaution. The precautionary principle is particularly relevant for the EU risk policy. It is applied by the European Commission in those specific circumstances where scientific evidence is insufficient, inconclusive or uncertain and there are indications through preliminary scientific evaluation that there are reasonable ground for concern that the potential dangerous effects on the environment, human, animal or plant health may be inconsistent with the chosen level of protection. Although the precautionary principle is mentioned in various international conventions and agreements, its application has been controversial in international trade disputes. Assessing risk management options involves the evaluation of risk-risk and risk-benefit balances, consideration of risk distribution, resolution of uncertainties and setting criteria for risk acceptability. Finally, effective risk management needs the establishment of transparent, participatory and open procedures in view of framing the appraisal of options within the wider societal perspective.

5.1 Introduction

In this Chapter we introduce the definition, the concept, contents, principles and practices of risk management. We stress again in this Chapter that while risk management and risk assessment are distinct functions, there is not, and there should not be, a rigid separation between the two. We focus on the approach to risk management within the EU decision-making process, where risk management is basically a preparatory process aimed at identifying, evaluating and selecting options for risk

B. Delogu, *Risk Analysis and Governance in EU Policy Making and Regulation*,
DOI 10.1007/978-3-319-30822-7_5

prevention, mitigation or response measures. Risk management is conceived as a technical process, aimed at preparing and supporting political decisions. On the EU level, risk management is framed within a comprehensive regulatory impact assessment process and is subject to the criteria and procedures that the European Commission has introduced with its Better Regulation initiatives. We summarise here the basic features of the EU decision-making process and discuss some general principles of the Treaties that are particularly relevant for risk policy and regulatory measures on the EU level. The Precautionary Principle is the basis for action in case of uncertainties. The Subsidiarity and the Proportionality Principles provide guidance for defining the scope and limits of the EU competence, notably of its involvement on risk matters, and the characteristic of action on the EU level.

This Chapter also examines some most important issues and criteria for risk management. Action may be necessary and justified on consideration of hazards, rather than risks, under particular circumstances. In the risk management practice it is often necessary to compare options for their risk-risk balance, for their risk and benefits and risk-cost balance, for the acceptability of the residual risks or the ancillary risks and the benefits of the measures considered.

Overall, the Chapter shows that risk management on the EU level is performed within a particularly complex framework, due to the multiplicity of institutional actors and the diversity of conditions and interests across the 28 EU Member States.

5.2 Risk Management and Risk Assessment

According to the classical risk analysis paradigm that we have introduced in Chap. 3, risk management, which includes the identification, assessment and selection of policy options, was traditionally presented as the second step of the analysis process, following risk assessment, in a simple linear sequence. For instance, the European Commission stated in its communication on Health and Food Safety of 1997 (European Commission 1997):

> Risk analysis is a three stage process composed of the assessment, the management and the communication of the risk concerned.

In fact, the two processes should be seen as deeply interconnected. In its seminal report of 1983 that we have already mentioned in Chap. 3 (US NRC 1983), the US National Research Council recommended "that regulatory agencies take steps to establish and maintaining a clear conceptual distinction between assessment of risks and consideration of risk management alternatives" but, at the same time stressed that although the distinction between risk assessment and risk management is important, it should not be taken to imply that they should be isolated from each other. It was noted that risk assessment and risk management in practice interact and that the necessary distinction between the two should not disrupt the desirable two-ways communication.

Moreover, the third component of risk analysis, risk communication, intended as a two-ways, interactive and participatory process, under current best practices runs in parallel to both risk assessment and management and involves assessors, managers, the relevant stakeholders and the public. For instance, EFSA, the EU agency in charge of risk assessment for food, has also been given a mandate to contribute to risk communication and has therefore established a communication strategy to that aim (EFSA 2012), noting that the European Food Safety Authority's mandate included as a key element communicating on risks associated with the food chain. In its communication strategy, EFSA stated that "by communicating on risks in an open and transparent way based on the independent scientific advice of its scientific expert panels, EFSA contributes to improving food safety in Europe and to building public confidence in the way risk is assessed".

Although good practices for risk analysis require that the independence of technical risk assessment is preserved from the influence of political, economic or other vested interests, collaboration between assessors and managers and openness to dialogue with and contribution from stakeholders and the public are necessary for ensuring the quality and utility of risk assessment. We will see in Chap. 8 that one of the main aspects for further improvement of risk analysis practices is the reinforcement of the dialogue between managers and assessors: the Commission Scientific Committees have recommended in a 2013 opinion (SCCS, SCHER, SCENIHR 2013) to establish:

> a system of dialogue that facilitates the exchange of information between risk assessors and risk managers while ensuring the scientific integrity of risk assessment ... extending this dialogue to all stakeholders in both initial forums and final consultations as a way of clarifying issues and ensuring more ownership.

Risk managers must provide inputs into the assessment process, in particular it is their responsibility to define and properly frame the issues to be addressed by the assessors. As we will see, risk managers need to interact with assessors at different stages, while respecting the integrity and independence of scientific risk assessment. Therefore, we state once again that the functional distinction between assessment and management of risks should not be misunderstood as a rigid separation.

Later in the book we discuss in some detail why the functioning of the interface between assessors, managers and political decision makers is vital for the effective risk policy making and how it could be improved. For now, we note once again that it would be a mistake to conceive risk assessment as a purely objective scientific activity, isolated from risk management and the relevant political, societal, legal and institutional framework. Among other considerations, assessors are themselves citizens, part of our societies: in framing issues and making judgements they may be influenced to a certain extent by their extra-scientific values, views and preferences.

Not all the steps and functions of risk assessment are mechanically determined by well-defined and codified scientific rules of the relevant disciplines. As we have highlighted in Chaps 2 and 4, room is left for subjective choice and judgement, notably for defining the scope of the assessment, deciding upon and framing the relevant aspects to be considered; for assigning subjective probabilities; deciding which evidence and data should be taken into account and which ones should be discarded; setting the strength of evidence required for concluding to a causal link between exposure to a stressor and a health effect, etc. Divergent opinions on the same issue by different groups of scientists may result from different assumptions and choices.

A striking example on how far and deep disagreement between scientists may be is offered by the endocrine disruptors controversy that we have described in Chap. 4. In some jurisdictions, guidance is developed and continuously improved with the objective to frame such judgments and choices. For example, the US National Research Council has devoted a Chapter of its report on "Science and Decisions, Advancing Risk Assessment" (US NRC 2009) to the selection and use of defaults. We will present further on in this book the guidelines established by EFSA and by the Commission Scientific Committees on how to weigh evidence. Nevertheless, a certain degree of subjectivity always remains. Assessment practices may be influenced by values, disciplinary background, adhesion to conventional schools of thought as well as personal preferences. Scientists involved in risk assessment may sometimes be pushed by their laudable ethical commitment to help policy makers in addressing and managing important health, environmental or safety problems, and by the expectations and requests of policy makers and the public, to provide assertive advice even when the available data and evidence are limited or uncertain, relying on professional judgement and subjective probabilities in the absence of robust evidence. An example is the imprudent statements by scientists and experts before the 2009 earthquake at L'Aquila (Italy) that killed more than 300 people. In spite of the impossibility of predicting earthquakes, but in order to reassure the population and help the authorities in preventing panic, the then deputy director of the Civil Protection agency, advised locals to sit back and enjoy "a nice glass of Montepulciano" wine. Six Italian scientists were finally cleared by an appeals court, but the court upheld a guilty verdict against the Civil Protection Director stating that while they accepted it was impossible to predict a quake, the experts had "minimised" the threat and created a false sense of safety in local inhabitants, with disastrous consequences.

As we will show later on, there is nothing wrong with the more subjective aspects of the role of assessors, provided that they are transparently explained, clearly understood by all the interested parties and the public and that a clear distinction is made between the purely "scientific" steps and the more subjective, uncertain or value-based assumptions or judgements.

Risk assessment policy guidelines should be established by risk managers in collaboration with assessors aimed at providing a consistent, recognised and transparent framework for the non-scientific steps of risk assessment. The absence of a transparent and stable frame for non-scientific aspects of risks assessment, established ex-ante, accepted and endorsed by both assessors and managers, may result in a too wide room for ad hoc adaptation and interpretation of risk assessment results. That may introduce inconsistencies and uncertainty in the way the assessment is performed by assessors and risks are addressed by managers and policy makers. There is a need for clear, stable, transparent and agreed "rules of the game". The aim of "risk assessment policies" is to provide such agreed frame for risk assessment. We discuss more in detail the aims and contents of risk assessment policy in Chap. 8.

5.3 The Concept and Contents of Risk Management

The standard ISO 31000:2009 (ISO 2009) on Risk management Principles and Guidelines defines risk management as the "coordinated activities to direct and control an organization with regard to risk". Risk is defined by ISO as the "effect of uncertainty on objectives".

The ISO definition is intended to apply across a wide range of organisations and different types of risks. However, organizations of different nature not only face substantially different risk-issues, but act within their own, specific legal, governance, management and operational frameworks. The very general definition provided by the ISO standard is therefore not sufficient to clarify the nature and contents of risk management activities in each specific organisation and policy area. In particular, within the EU decision making process, risk management is very precisely understood to include the identification of policy options directed at preventing, mitigating or responding to the risks addressed, taking into account the results of risk assessment and the requirements of the applicable legal framework, and the assessment of the options so identified for their effectiveness and efficiency, cost and benefits, social impacts and other relevant implications, with a view to achieve the desired level of protection.

The European Commission stated in its mentioned communication on Consumer Health and Food Safety that

> risk management shall include the process of assessing the impact of policy alternatives in the light of the results of risk assessment and the desired level of protection.

The US-NRC "Red Book" defined it as a process of weighing policy alternatives. The objective is to select the most appropriate regulatory action and the approach involves integrating the results of risk assessment with engineering data and with social, economic and political concerns to reach a decision.

This "Red Book" definition goes further than the definition of the European Commission by including under risk management consideration of "political concern", which in principle are reserved in the EU to the final political appraisal and decision-making stage.

In light of these definitions, risk management should be considered to include several functions, which apply all along the EU assessment-management-decision process, as described in Table 5.1:

Table 5.1 Functions of risk management and risk managers

1. To identify and define the risk issue to be addressed
2. To frame the risk issue identified, in relation to:
The relevant legal aspects
The factual aspects that, on the EU level, include in particular the situation of the member states and their assessment of risk and management measures
The ethical and value-related issues
The economic, social and trade aspects
The stakeholder interests, values, objectives, positions and perceptions
3. To preliminarily define the broad policy objectives and options
4. To organise and initiate the risk assessment process:
To identify the appropriate body for providing advice
To ensure the availability of appropriate independent expertise
To discuss and define a mandate for risk assessors
To ensure access to data and information needed for the assessment
5. To maintain a sustained dialogue with risk assessors across the risk assessment process:
To clarify, revise and complete the mandate as it may be necessary
To ensure an agreed approach on risk assessment policy issues (defaults, risk quantification criteria etc.)
To obtain clarification on, and gain full understanding of the risk assessment results and conclusions
To monitor compliance with criteria on quality, independency and transparency
6. To analyse risk assessment results, resolve uncertainties and appraise the risk in relation to management criteria
7. To identify the policy/regulatory options to be considered
8. To assess the policy options (notably in light of the applicable impact assessment guidelines)
9. To ensure transparency, consultation and participation across the various steps of the process, in collaboration with the risk assessment body:
To explore stakeholder views, interests, values, objectives, positions and perceptions on risk at an early stage
To consult widely on the draft risk assessment mandate
To ensure public availability of information on the process
To organise stakeholder or public consultations on relevant data and information
To organise public consultations on the draft risk assessment opinions and ensure an appropriate follow up and feedback to comments received
To consult widely on the policy options considered
To inform on the conclusions reached and their rationale
10. To present conclusions to the decision makers, facilitate dialogue between decision makers and scientists and support the decision making process

Although some of these functions are often delegated to bodies or services ensuring technical and administrative support to risk assessors, risk managers are responsible for ensuring that the appropriate structures, procedures and resources are in place and that they function properly. For instance, EFSA, the body providing advice on food risks to the EU Institution, selects the experts sitting in its scientific committee, panels and groups, organises public consultations and contributes to risk communication. In doing that, EFSA is independent from the Institutions, notably from the Commission, but its mandate, rules and principles of functioning have been decided by the European Parliament and the Council on proposal from the Commission. According to the EU case law, the Institutions are in fact responsible for the quality and reliability of the advice on which they base their decisions.

Some further considerations help clarifying the role and nature of risk management as part of the EU decision-making process.

First, risk management is distinct from political decision making. Risk management is part of the preparatory process leading to decisions by the EU political institutions. It is performed at technical, administrative level, and involves, as mentioned above, an evaluation of the policy options, in light of the available scientific knowledge, for their effectiveness (in relation to the protection objectives and the desired level of protection), their cost-efficiency, feasibility, fairness, conformity with the applicable legal requirements and principles, including on ethical and social aspects, and their compatibility with international trade rules.

Second, it is the risk managers' responsibility to define the risk issue being considered, set its boundaries, clarify the context of the issue and the objective being sought, to formulate the questions to be submitted to risk assessors, to examine the assessment results, analyse their relevance, possible limitations and their policy implications, and resolve the uncertainties highlighted by risk assessors. These functions require a sustained dialogue with the risk assessors. Since risk management is intended to provide the basis for political decisions, risk managers must ensure to properly address decision makers' needs.

Risk management is therefore a technical function. Decisions are taken by the EU political institutions, the Parliament and the Council, on proposal from the Commission. Certain EU decisions are delegated by the European Parliament and the Council to the Commission, under well-defined conditions. Those decisions are prepared by Commission services, but are decided upon by the Commission acting on its political level, the "College" of Commissioners.

It is indeed important to underline that decisions on risk may not mechanically result from a purely technical process. Once again, the final, decisive political appraisal and decision take always place in the EU on the political level, in compliance with the democratic principle of accountability that prevents attributing political decision powers to scientists, experts or civil servants. The final political appraisal is the way in which societal preferences are brought into play in a democratic legal order. Therefore, the important objective of "science-based policies" should not be misinterpreted as an attempt to frustrate the right of the public to determine important decisions on their health and safety and on the environment, through their political representatives.

5.4 The EU Risk Managers and the Regulatory Decision Process

We now need to say a few words about the procedural aspects of the EU decision-making process, the nature of the legal acts through which decisions on risk management measures are established and on the intervening institutional actors. The EU could not be compared to a national state or a federation. Its peculiar nature is a source of complexity in the decision-making process that has a significant impact on the way in which the risk management principles are applied. We have just highlighted that on the EU level, policy and regulatory decisions are prepared and proposed by the Commission. We have seen that the preparatory work is conducted by Commission services, which play in that respect the role of risk manager and normally rely upon the scientific support of dedicated, independent risk assessment bodies as explained in other parts of this book. The Commission has established very specific and detailed internal rules on the procedures applicable to the preparatory work by its services, notably with its Better Regulation initiatives, which are described and discussed further on in this Chapter and in Chap. 7. The final, political step of the process, the decision taken on the political level by the "college" of Commissioners, intervenes both for the legislative proposals presented for decision to the Parliament and the Council and for the measures adopted by the Commission itself, based on its implementation or delegated powers.

Commission services do not act as risk manager in isolation. Close links are often established with the Member States through committees or working groups of national experts that bring in the expertise, knowledge and positions from across the EU. Important examples are the committees assisting the Commission on food safety, chemicals, pharmaceutical products and cosmetics.

There are different types of EU Committees and permanent or ad hoc consultative groups assisting the European Commission in the preparation of proposals or the adoption of delegated and implementation measures. Some of the committees are set up by legislative acts and have an institutional role in the decision-making process under the so-called "comitology". Through comitology procedures, the EU Member States contribute to, and control the implementation of EU law by the European Commission. The Commission is bound to consult, before deciding upon an implementing act, a committee composed of representatives of the EU Member States. Such committees provide opinions on the Commission's proposed measures, which are more or less binding, depending on the relevant provisions in the legal act being implemented.

For example, Regulation 178/2002 on food law and food safety (EU 2002) set up a Standing Committee on Plants, Animals, Food and Feed with a mandate covering the entire food supply chain. The Committee has an important function, well beyond the "control" on the Commission measures, helping the EU to deal effectively with health risks at every stage of the production chain. It is organized in sections, is composed of representatives of the specialized agencies and ministries of the member States, and is chaired by the Commission itself.

Moreover, risk issues are often raised on the EU level by the Parliament or the Council, or result from developments in the Member States. And Commission consultation procedures ensure a substantial input from stakeholders along the risk management process on all major legislative proposals. Commission consultation procedures are being reinforced and extended to cover most of Commission decision under the recent "Better Regulation Package".

Within the Commission services, a wide range of profiles are available, ensuring in-house legal, economic and technical expertise and support for the risk management functions. In addition, Commission services are often supported by external, specialised consultants when the internal capacity is not sufficient to deal with particular subjects or with the workload resulting from demanding functions, like in particular extensive collection and analysis of data.

The EU Decision Making Process The EU is a very peculiar entity. It is not a federal super-state, but it is not an intergovernmental organisation either. As a sort of "federation of nation-states", according to the definition given by the former Commission President Jacques Delors, it has an articulated distribution of competences between the EU and the national levels and complex institutional structure and decision making procedures, reflecting both its supranational nature and the role of its sovereign Member States. The decision making procedure that applies to policies and measures related to health, safety and environmental risks is the "ordinary legislative procedure". It is useful to describe here the EU legislative procedure in order to show how complex is the application of risk-analysis best practices to legislative measures on the EU level.

The Ordinary Legislative Procedure includes several steps:

- Proposal: the Commission submits a proposal to the European Parliament and the Council. The Commission has the exclusive right to make proposals: in the policy areas considered, the Parliament and the Council may only take decisions based on a proposal from the Commission. The Council may amend a proposal from the Commission only acting unanimously.
- First reading: the Parliament adopts a position on the proposal and communicates it to the Council. Generally, the Parliament proposes amendments at this stage.
- If the Council approves the position of the Parliament, the act is adopted as amended by the Parliament. Otherwise, the Council adopts its position and communicates it to the Parliament. The Commission informs the Parliament of its position.
- Second reading: within 3 months from the communication of the Council's position, the Parliament:
  - Approves the Council's position or does not take a decision: the act is adopted in the wording of the Council;
  - Rejects the Council's position: the act is not adopted;

  - Proposes amendments to the Council's position and communicates it to the Council and the Commission. The Commission deliver an opinion on the proposed amendments.
- Within 3 months of receiving the Parliament's amendments, the Council:
  - Approves all the amendments: the act is adopted in the wording of the Parliament;
  - Does not approve all the amendments: a Conciliation Committee between the Parliament and the Council with the participation of the Commission is convened.
- Conciliation: within 6 weeks from its being convened, the Conciliation Committee:
  - Reach agreement on a joint text based on the second reading positions of the Parliament and the Council: the joint text is submitted to the Parliament and the Council for a third reading;
  - Does not reach agreement on a joint text: the act is not adopted.
- Third reading:
  - The Parliament at the majority of voted cast and the Council at the qualified majority approve the joint text: the act is adopted
  - The Parliament and the Council fail to approve the joint text: the act is not adopted.

This brief description of the relevant decision making procedure shows that a multiplicity of actors intervene in the EU decision-chain at various stages of the process. A deeper insight into the internal dynamics of the Institutions would show how the diverse political orientations and national interests further complicate the search for agreement on risk-protection measures. Under such conditions, ensuring a coherent application, along the entire process, of best practices for risk analysis, participatory governance and the regulatory approach is a challenging objective. In practice, it is mainly on the level of the preparatory work by the Commission, that the independent scientific input is organised and brought into the decision process, stakeholders are widely and transparently consulted and a systematic assessment of options is performed.

Progress has been made in applying impact assessment provisions to amendments introduced by the Parliament. As already mentioned, the Commission has recently proposed a reinforcement of Better Regulation, including in particular its impact assessment practices, across the entire institutional decision-making process (we say more on this initiative in the next section). There are however practical, legal and political constraints limiting the introduction of binding procedural requirements applicable to the decision making powers of the Parliament and the Council, beyond what is foreseen in the Treaties.

As mentioned before, many relevant decisions are however delegated or left to the Commission. Delegated and implementing measures have a particular importance for EU risk policy in the health, safety and environmental areas. Many standards, authorisations, positive and negative lists of substances etc. are decided through such implementing or delegated acts (briefly described here below). For instance, the annexes to the Cosmetics Regulation may be amended by the Commission assisted by the Standing Committee on Cosmetic Products. Those measures, designed to amend non-essential elements of the Cosmetic Regulation, are adopted in accordance with a regulatory procedure with Parliamentary scrutiny. Such measures have in fact substantive risk management significance as they include among others the updating of lists of substances prohibited or subject to conditions for use in cosmetics. Other examples are the measures authorizing the deliberate release of GMOs or authorizing use of chemicals of high concern in particular applications. It is easier in such very relevant risk management cases to implement best practices consistently and systematically because a single actor, the Commission, has a dominant role in the applicable decision process.

The Delegated and Implementing Acts A legislative act may empower the Commission to adopt non-legislative acts of a general application to supplement or amend certain non-essential aspect of the legislative act. The TFEU states that the legislative act must establish the conditions applicable to the delegation, including in particular the possibility for the Parliament or the Council to revoke the delegation or to object to a delegated act preventing it to enter into force.

As a general rule, implementation of the EU legislative acts falls under the responsibility of the Member States. Nevertheless, where uniform conditions are needed for the implementation of a legislative measure, implementing powers may be conferred by the legal act on the Commission. The exercise of implementing powers is subject to general principles set by the Parliament and the Council, concerning mechanisms for control by the Member States.

Finally, the EU regulatory policy is undergoing a process of scrutiny and likely revision toward more diversified approaches that may attribute a greater role to non-institutional actors. We say more on the EU regulatory policy and its evolution in Chap. 7.

5.5 Risk Assessment, Risk Management and Impact Assessment (IA)

Risk management as a preparatory stage of EU decision making is implemented in accordance with the procedural and substantive requirements of the specific Treaty provisions or secondary legal framework applicable. In the case of legislative proposals, risk management as a process to evaluate, compare and choose policy options is subsumed under the EU Impact Assessment process that also defines the role of risk assessment. We have said that the scope of application of Impact Assessment is being extended by the European Commission to cover delegated and implementing measures of significant impact.

Formal and substantive requirements applicable to certain types of measures implementing general legislation are usually precisely defined by the relevant legal framework. In such cases, the decision procedure to be followed and the type, form and content of the legal instruments to be used are determined by the relevant legal provisions. There is very little room for policy or regulatory options deviating from the procedural provisions and decision criteria of the applicable legislation. Examples are the measures authorising the placing on the EU market of medicines, under the EU legislation on pharmaceutical products, and those authorising specific applications of substances of high concern (for example carcinogenic, mutagenic or repro-toxic substances) under the REACH regulation.

Apart from implementing or delegated acts with a narrow scope, routine or urgent character, for the wider range of EU policy or regulatory proposals aimed at health safety and environmental protection risk management is subject to the impact assessment procedure that the European Commission applies as part of its "better regulation" policy (European Commission 2006).

The Better Regulation principles and the impact assessment (IA) procedure, which is a key element of better regulation, have been endorsed by the Council and the Parliament, which are also committed to consider the impact assessment of the proposals provided by the Commission. The Parliament has established its own impact assessment procedure to assess its amendments to the Commission proposals, which however so far has been applied only in a limited number of cases, while the Council so far has not implemented an assessment of its amendments. A recent Commission initiative (presented in May 2015) (European Commission 2015-1) is aimed at reinforcing, by a new inter-institutional agreement, the role of impact assessment across all the stages of the EU decision process.

The Commission established a set of guidelines (European Commission 2009-2) to ensure the quality and consistency of its impact assessments. The Commission guidelines are revised when necessary in light of experience and a new set of comprehensive Better Regulation Guidelines, including guidance on Impact Assessment replacing the narrower 2009 version, has been adopted by the Commission in May 2015 (European Commission 2015-2). The Commission has also developed a wide range of specific assessment tools, notably for assessing costs and benefits, the administrative burden and the impact on SMEs.

The Better Regulation rules define the procedure to be followed for planning, developing and assessing legislative initiatives and other proposed decisions. They also include procedures for the consultation of stakeholders and for ensuring transparency across the entire preparatory process.

Three broad aspects are addressed in the course of the EU regulatory impact assessment: economic, social and environmental impacts. Environmental impacts are defined in a comprehensive way and include, in addition to ecological aspects, human health and animal health and welfare. Particular attention is devoted to the subsidiarity and proportionality aspects, according to principles that we will discuss further on.

On risk assessment, the 2009 IA Guidelines stated that

an IA that addresses a problem in which uncertainty about serious negative outcomes (risks) is an issue should contain a risk assessment. If these risks may involve irreversible damage or fatalities on an unforeseeable scale, a separate formal risk assessment will have to be carried out on the basis of scientific expertise.

The Commission impact assessment guidelines recognise the concept of risk as a function of the probability and severity of an adverse outcome. We have seen in Chap. 2 how the concept of risk is defined in the Toolbox document accompanying the new (2015) IA Guidelines. The 2009 guidelines on when and how to carry out a risk assessment did not go into the details of risk analysis as internationally recognised. In practice, risk assessment was and is however conducted by dedicated, independent bodies (notably EFSA, ECHA, EMA, ECDC and the Commission non-food Scientific Committees) that apply well-established assessment procedures and methods, as presented in this book.

The Toolbox document accompanying the new 2015 Guidelines provides more specific and detailed guidance on risk assessment and management. In addition to introducing the already mentioned definition of risk, the document refers to the key role of the EU specialised decentralised Agencies and the Commission Scientific Committees for providing the necessary scientific support, account taken of the required in-depth expertise and specialist knowledge, spanning various risk and policy areas, needed for assessing complex and diverse risk matters.

Under the current EU practice, the scope for the application of risk assessment is wider than that of IA since risk assessment is required also for case-by-case Commission decisions, like those implementing EU food safety legislation or those taken under the REACH Regulation aimed to the restriction of individual chemicals, where so far a IA has in general not been applied (but in the case of REACH any restriction or authorisation requirement is also assessed for its socio-economic aspect by the Socio-Economic Assessment Committee of ECHA). Nevertheless, the Toolbox document states that an IA should be established in cases where the Precautionary Principle is invoked, or the decision deviates from the conclusions of the assessors. Since under the EU risk analysis paradigm assessors are not expected to provide risk management recommendations, this last provision should be understood as referring to cases where the overall appraisal by risk managers of the need for action deviates from the risk assessment conclusions.

The Joint Research Centre of the European Commission complements the expertise of the risk assessment bodies and committees where necessary. It may provide tools and methods for risk assessment and validate risk assessment methodologies. It may also provide advice in case of divergent opinions or issues characterised by high uncertainty. The Toolbox document also briefly refers to the classical steps of risk assessment and discusses the various types of uncertainties that may affect risk assessment and must be evaluated as far as possible and transparently reported.

Moreover, the document discusses risk tolerability and risk criteria, noting that such criteria may be included in the legislation or derive from scientific knowledge of tolerability thresholds. Risk controllability, risk-benefit trade-offs, risk perception, societal values are recognised as relevant and legitimate issues to be considered when establishing risk criteria.

The Concept of Tolerable and Intolerable Risk According to the European Commission An *intolerable risk* is so significant that risk management measures should be taken to eliminate the hazard and/or the exposure. However, it should be noted that the elimination of one risk, for example by banning a particular hazardous chemical, could result in its replacement by another, potentially more significant and uncertain risk (i.e. substance with unknown effects on human health). Where it is not possible to eliminate an intolerable risk (e.g. in the case of natural hazards), it should be at least be reduced by mitigation and preparedness measures.

A *tolerable risk* may be worth reducing through actions by private and/or public actors. Even where there is no, or just a negligible risk (sometimes also called "acceptable" risk), there could be reasons for public or private intervention (e.g. on a voluntary basis). Public perception of a risk may for example require an effective risk communication strategy.

On risk management, the Toolbox document refers to the need of considering the most proportionate measures and to the possibility of acting on the hazard and/or on the likelihood of the hazard to materialise. The assessment of options should compare the marginal costs and the marginal benefits of risk reduction or, when such marginal costs and benefits cannot be calculated, the total costs and benefits. Nevertheless, it is recognised that it is not always possible to quantify risk reduction benefits in monetary terms

> there may be impacts and/or likelihoods that are not possible or appropriate to quantify but that should be taken into account nevertheless (e.g. where robust monetary values are not readily available as in the area of security, freedom and biodiversity or where the high level of uncertainty renders any quantification meaningless.

Finally the document identifies the contributions of risk assessment to the IA process as summarized in Table 5.2.

While this table includes the main technical elements of the relationship between risk assessment and IA in the purely internal Commission process, it does not mention the wider role of risk managers and the other important elements of best practice like in particular stakeholder input and multi-dimensional risk appraisal. Relevant additional elements in that respect are however included under other provisions of Better Regulation.

The interface between risk assessment and impact assessment requires a careful conceptual and practical coordination. The aim of impact assessment is to quantify as much as possible the pros and cons of the different options identified, and to evaluate them, as much as feasible, in quantitative terms and with consistent metrics. However, the results of the scientific risk assessment are often presented in technical terms which cannot be easily translated into economic quantities.

Table 5.2 Risk assessment and impact assessment

	Risk assessment	IA process	Main actor(s)
1.	Identify potentially significant risks	Identify problems	Lead DG together with ISG (with input from risk assessors where relevant)[a]
	Identify how and by whom the risk assessment will be carried out		
2.	Assess risk and uncertainty	Assess problems and baseline	Risk assessors
	Complement and/or validate the risk assessment if needed		
3.	Identify risk criteria and evaluate risk	Define objectives	Lead DG together with ISG (with input from risk assessors where needed)
4.	Develop risk management options to eliminate, transfer of reduce risk	Develop options	
5.	Use risk assessment to assess impacts, use sensitivity auditing to assess uncertainty	Assess options	
6.	Plan for communicating risk, reducing uncertainty, adapting the risk management approach if necessary, monitoring new and existing risks etc.	Outline monitoring and evaluation arrangements	

[a]DG means a Commission Directorate General, while ISG refers to the Inter-service Group of DGs associated to the development of the IA

Moreover, defining policy objectives and criteria for human health (like for example deciding the objectives of protection measures based on criteria for acceptable levels of cancer risk) or the environment (for example establishing a nature-preservation or a land-development policy based on monetary values attributed to biodiversity or the landscape) may pose difficult ethical and political problems. Certain methods and criteria exist and are used in practice. They produce useful indicative results, but have not been endorsed on the EU political level and therefore do not carry much weight in the political debate. Such issues are unlikely to be settled in general terms on the political level. Furthermore, risk assessment is often characterised by uncertainties that make it impossible to compare policy options and identify the "best" one in a deterministic way. The Commission IA Guidelines therefore refer to the need to apply the precautionary principle (PP), as appropriate, in case of uncertainty. The PP, the next subject of this Chapter, properly applies under the specific conditions that have been defined by the Commission in its communication of 2000 (European Commission 2000). It legitimates proportionate action to prevent risks when a full risk assessment is not feasible.

The optimisation of the interface between risk analysis and impact assessment based on experience deserves some further considerations and will therefore be further discussed in Chap. 8 on the improvement of the utility of risk assessment.

In conclusion, risk management on the EU level takes place within the standard processes applicable to the preparation of policy and regulatory measures and is aimed to support the EU institutional decision making procedures. Differently from other jurisdictions, there are no specialised agencies in the EU with specific regulatory competence for the different risk-sectors. There are no specific procedures for risk management as such either. Risk management in the EU decision-making process is characterised by the key support-role assigned to risk assessment, the application of regulatory impact assessment procedures to management options and by some specifically relevant principles, criteria and issues.

Some of the considerations presented in the following sections anticipate on our discussion on the characteristics, issues, problems and perspectives of the EU risk regulatory policy. We will discuss EU Better Regulation and Impact Assessment in a wider and deeper perspective in Chap. 7.

5.6 Risk Management Principles and Criteria

Risk management in the EU is guided by a few overarching general principles that are established by the EU Treaties and incorporated into secondary law, and on specific guidelines and practices. The principles most relevant for the EU action on health and environmental protection are the Precautionary Principle (PP) and the Principles of Subsidiarity and Proportionality.

5.6.1 The Precautionary Principle (PP)

Introduction The Precautionary Principle is often of particular relevance for guiding risk-decisions since, as we have discussed in previous Chapters, risk assessment is frequently affected by important uncertainties of various nature. The dictionary definitions of precaution are not very useful for understanding what is addressed under the Precautionary Principle. For instance, the Webster Dictionary defines precaution as "a measure taken beforehand against possible danger". This meaning is a bit ambiguous as it does not expressly refer to the uncertain of the outcome, which is an intrinsic characteristic of risk, and therefore may be understood as addressing preventive measures on deterministically determined adverse events, measures taken before a danger materialises. However, we are all in fact very familiar with the concept of precaution in deciding what course of action or behaviour to take in case of uncertainty about possible adverse events or outcomes. Symmetrically and concurrently with the preparedness to take reasoned risks in view of potential and proportionate expected benefits, human ability to act with precaution is probably one of the reasons for the success of humankind in the evolutionary selection process. Similarly to pain, which is a signal helping to preserve the integrity of an organism, fear and the psychological and rational processes inducing precaution help preserving safety, health and preventing losses. In the layman wisdom, lack of precaution is considered a sign of irresponsibility, immaturity, irrationality.

In conceptualizing precaution, there is no great difference between action by individuals and collective action mediated by political decisions. Requiring that an additive be tested for possible toxicity before being allowed into marketed food in order to prevent the risk of health damages is exactly as rational as deciding to check the weather forecasts before booking a holiday at a seaside resort in order to prevent the risk to waste money and days of leave from work. And throwing away a soup that has remained in the refrigerator for several days and may or may not be any longer good for consumption is as rational as not allowing nanoparticles in food unless they have been demonstrated to be safe for consumption.

Simple examples like these refer to an intuitive application of the Precautionary Principle. Still, there has been, and there still is much confusion and controversy about the PP. Although precaution is accepted and praised as a rational principle of conduct for individuals' decisions and behaviour, for instance when driving on a highway, it is sometimes criticised as driving societies to inertia and ruin when assumed as a principle for public policies. For example, in an article published in "Regulation" Cass R. Sunstain of the University of Chicago warned that (Sunstein 2002) in his opinion the problem with the precautionary principle was not that it leads to wrong decision, but that, if strictly interpreted and applied, it would paralyze decision-making.

In fact, the positive or negative impact of the precaution depends entirely on the way in which the principle is defined and put into practice and on the resulting balance between precaution and risk taking. John Graham, a former Administrator of the Office of Information and Regulatory Affairs at the US Office of Management and Budget, wisely noted in a speech to the annual meeting of the Society for Risk Analysis in 2002 that "the use of precaution in risk management is sensible but susceptible of misuse".

Again, it goes in public policy as in individual choices: excess of precaution can lead to extravagant individual behaviour (taking the umbrella in a sunny day) as well as to counterproductive and penalising measures (restricting a technology in the absence of any indication of a plausible mechanism for health or environmental hazard). Risk assessors and decision makers are frequently faced with uncertainty when dealing with risk issues. As explained in Chap. 4, risk assessment is fraught with uncertainties and limitations at all stages. The first challenge in case of new products, chemicals or technologies is to identify all their hazards. Without appropriate testing or other forms of systematic examination, it is often impossible to say whether or not a new product, substance or technology presents a hazard. This may bring the authorities to establish a "precautionary" requirement to test in order to check at least the most plausible hypotheses as a preventive risk management measure. Moreover, once certain hazards have been identified, there may still be uncertainty on the dose-effect relationships, the validity of extrapolation to humans of toxicity data from animal tests, and/or about the level of exposure.

Uncertainties, limitations to scientific knowledge or lack of data do not exonerate decision makers from the responsibility to choose a course of action. Deciding not to decide, i.e. not to take action or to postpone decisions, involves political and legal responsibilities in case unforeseen adverse impacts materialise. Uncertainty does not lift political accountability nor, depending on circumstances, legal responsibility for damages. There is unfortunately plenty of examples of the disastrous consequences and the heavy political and legal responsibilities related to the failure to act effectively on early warnings. It suffice to mention the well-known stories of asbestos, contaminated blood in France and the BSE crisis (See for example: "Late lessons from early warnings: science, precaution, innovation", EEA Report No 1/2013).

The Precautionary Principle (PP) established by EU Treaties is aimed at guiding EU Institutions in their decisions on health and environmental risks in cases of uncertainty or incomplete knowledge.

In applying the PP, decision makers have often to strike a difficult balance between on the one hand the right to enterprise and trade, as well as the needs of innovation with its potential benefits and, on the other hand, the need to ensure effective, high level health and environmental protection. Divergent interests, views, political stances push into opposite directions. Economic and industrial lobbies often plead for "more research" and for acting only if and when robust evidence becomes available. Other civil society organisations push instead for immediate precautionary action. Depending on their interests or political agendas interested parties are induced to either interpret in an excessively extensive way the PP or to criticise it as being just a pretext for political expediency or protectionism. The arguments on both sides are often specious. A great degree of objectivity and a good, shared understanding of the Principle are therefore required in order to dispel the misinterpretations of the PP. This part of Chap. 5 aims to provide some information and considerations that may help in that respect.

A few general considerations should be made here on the controversy around the PP.

The PP is sometimes invoked for requesting decision makers to take action on purely hypothetical risks. In that respect, as we will show in more detail, the EU Court of Justice has clearly stated that a precautionary measure may not be properly based on a purely hypothetical approach to risk. For instance, the European Commission has invested in promoting and monitoring scientific knowledge on the effects of low frequency radiation on human health, supporting research to possibly identify any "unknown unknowns". It has however resisted pressure to take immediate action, based on claims of hypothetical hazards, beyond the enactment, through product emission standards, of the existing internationally recognised exposure limit values protecting against the proven ones.

At the opposite end, measures properly based on the PP are sometimes resisted by economic actors on grounds of absence of clear evidence of risk, an implicit request for certainty that is clearly incompatible with the very definition of the PP, which is a principle of action in case of uncertainty or lack of sufficient knowledge or data, conditions under which a full risk assessment is obviously not feasible. For instance, although already by mid of the 1970s there were relevant indications of the accumulation of CFCs in the stratosphere and laboratory tests had shown the ozone-depleting properties of Chlorine released by CFCs, industry resisted precautionary regulatory action in the absence of a more relevant assessment of the depletion risks based on direct evidence from atmospheric observations, which of course were not available at the time.

As we will further discuss later on, there are two typical situations in which the PP is respectively challenged or misused:

- The PP is criticised and opposed by those who claim that measures may only be legitimately taken on the basis of "sound science". That would require a full risk assessment showing without uncertainty the existence of an unacceptable risk. The interests defending this approach know well that there is no such thing as a "sound science of risk" and that risk assessment of complex and controversial issues is most often exposed to uncertainty and open to challenge.

  Interests opposed to a risk-prevention measure often strive to produce doubts and highlight uncertainties each time that a risk assessment study shows the plausibility of a risk. The PP is a challenge to potentially endless delaying strategies, as it allows action even when uncertainty prevents a full demonstration of the risk in the short term. Talking examples of the delaying strategies based on such "sound science" paradigm are the tobacco industry challenges to the health effects of smoking and the well alive "industry of doubt" on the reality and causes of climate change. The tobacco industry even established in 1993 a "sound science" lobby, the Advancement of Sound Science Center, formerly "The Advancement of Sound Science Coalition", to promote research producing inconclusive results in order to spread doubts on the scientific findings on the adverse health effects of smoking, and publicise articles and positions of scientists favourable to industry's arguments. The strategic objective in cases like these is gaining time by playing the "sound science" card, in spite of the implausibility of the criticisms and the hopelessness of winning the argument in the long term.
- In the opposite direction, the PP is sometimes stretched to justify with weak claims of hypothetical risks restrictive measures, like bans on certain products or technologies, which are invoked by the public opinion or by certain stakeholders mainly for reasons and preferences not really related to health or the environment and that do not provide a solid legal ground, under the current criteria of the EU legal order and the international trade rules, for creating barriers to trade. These situations pose the more fundamental issue of democratic sovereignty, i.e. whether a society is entitled to enforce through its democratic decision-making mechanisms its societal values and preferences and to protect its modes of production resulting from policy and regulatory choices. There is little clarity on the "legitimate requirements", in addition to the internationally recognised right to health and environmental protection, that may justify measures creating barriers to trade. International trade agreements have introduced a discipline limiting the scope for unilateral measures, although the public may not have been made fully aware of the scope and implications of such limitations. Such discipline creates a tension between the political responsibility of democratic institutions that must take account of the will of their constituencies and the external constraints limiting the freedom of political institutions to act. In such situations, policy makers, in order to defend measures aimed to respond to public preferences, may sometimes be tempted of focusing their line of defence on the risk management

dimension, invoking the PP even when the plausibility of a risk is weak. Although such measures are most often not intended to pursue protectionist objectives, they may have a significant impact on trade and therefore invoking the PP under such weak bases creates international trade conflicts and affects the credibility of the principle.

The Precautionary Principle as a Principle of EU Law Article 191 of the TFEU establishes the Union policy on the environment and states, among its various objectives, that it shall contribute to protecting human health. The same Article disposes that Union environmental policy shall be based, inter alia, on the precautionary principle.

The Treaty does not provide a definition of the PP. Nevertheless, the EU Court has clarified the nature and scope of the PP in its case law pronouncements.

According to the Treaty and the Court's doctrine, the Precautionary Principle applies to measures taken for the objectives of environmental policy. It also applies where the Union institutions take measures to protect human health in the framework of other Union policies. In fact, the Treaty states that environmental protection requirements must be integrated into the definition and implementation of the Union's policies and activities. Therefore the PP, which is part of the environmental policy principles, may be applied across all EU policies, as appropriate, if health or environmental issues are at stake.

In a judgement on the validity of the Commission's decision banning the exportation of beef from the United Kingdom to reduce the risk of BSE transmission (European Court of Justice 1998), the Court confirmed the legitimacy of precautionary measures and stated that:

> That approach is borne out by Article 130r(1) of the EC Treaty, according to which Community policy on the environment is to pursue the objective inter alia of protecting human health. Article 130r(2) provides that that policy is to aim at a high level of protection and is to be based in particular on the principles that preventive action should be taken and that environmental protection requirements must be integrated into the definition and implementation of other Community policies.

The Court has pronounced on the interpretation and application of the precautionary principles in several cases. It has stated that where there is uncertainty as to the existence or extent of risks to human health the Union institutions may, by reason of the precautionary principle, take protective measures without having to wait until the reality and seriousness of those risks become fully apparent.

The Court has further clarified that in a situation in which the precautionary principle is applied, which by definition coincides with a situation in which there is scientific uncertainty, a risk assessment cannot be required to provide Union institutions with conclusive scientific evidence of the reality of the risk and the seriousness of the potential adverse effects, were that risk to become a reality.

At the same time, the Court has set the limits of precaution by stating that a preventive measure cannot properly be based on a purely hypothetical approach to the risk founded on mere conjecture which has not been scientifically verified.

In other words, it can be concluded that:

- The precautionary principle is a general principle of EU law that applies across all the EU policy areas to measures that have environmental and health protection objectives or implications;
- The precautionary principle applies when, because of the existing scientific uncertainty, risk assessment cannot provide conclusive scientific evidence of the reality of a risk and of the seriousness of the potential adverse effects;
- The precautionary principle may not be invoked to take measures based on a purely hypothetical approach to risk, founded on mere conjecture, where the plausibility of such conjecture has not been scientifically verified. Although it is true that no evidence of risk is not evidence of no risk, groundless hypotheses may not be considered as valid reasons for precautionary action.

It should be noted that the PP allows measures to be taken under the said conditions, but does not dictate the nature of the measures. Risk management options must be evaluated in light also of the other relevant principles of EU law, such as subsidiarity and proportionality, as well as other relevant considerations as costs, benefits, ethical aspects etc.

Finally, a clear distinction is made within the EU risk analysis approach between the Precautionary Principle, which is a risk management principle guiding the assessment of policy and regulatory options under the specific conditions applicable to it, and certain precautionary assumptions made by risk assessors, for instance worst case assumptions for calculating exposure or the application of safety factors when extrapolating toxicity data from animals to humans. In fact there is no agreement on whether the application of safety/uncertainty factors could be considered "precautionary" since such factors are intended to address real differences and variability in the *inter-species* and *intra-species* response to stress factors.

The Commission guidelines on the Precautionary Principle The Commission has established in the year 2000 guidelines on the way it interprets and intends to apply the Precautionary Principle. These guidelines, in the form of a Communication, have three main objectives: support the Commission itself in its role as risk manager; take a position on the subject vis a vis the major trading partners and the international agencies; clarify with the Member States the conditions and limits for invoking the PP to justify national measures creating barriers to EU internal trade.

The Commission guidelines have been endorsed politically by the European Council. They have no binding effect, but the Court, pronouncing on a disputed measure on the withdrawal of the authorisation of an additive in feeding stuffs, has clarified that

> the Community (*now the Union*) institutions may lay down for themselves guidelines for the exercise of their discretionary powers…in particular by communications…In such circumstances the Community judicature ascertain, applying the principle of equal treatment, whether the disputed measure is consistent with the guidelines that the institutions have laid down for themselves by adopting and publishing such communications.

Therefore, the Commission guidelines are more than a mere announcement of intentions and have a legal useful effect as confirmed by the Court case law.The Communication does not provide a formal definition of the PP, but states that it covers:

> those specific circumstances where scientific evidence is insufficient, inconclusive or uncertain and there are indications through preliminary objective scientific evaluation that there are reasonable grounds for concern that the potential dangerous effects on the environment, human, animal or plant health may be inconsistent with the chosen level of protection.

This sentence includes the elements characterising Commission views on action under the PP:

- The PP is not an alternative to risk assessment. It may be invoked as a basis for action when it is not possible to perform a full scientific assessment of the risk because the available evidence is insufficient, inconclusive or uncertain;
- The PP is applied following an objective preliminary scientific evaluation;
- The evaluation must provide reasonable ground for concern of potential dangerous effects;
- Concerns of potential dangerous effects are related to the chosen level of protection.

It is for the EU institutions to choose the desired level of protection. It is internationally recognised, for instance in the Agreement on Sanitary and Phytosanitary measures (the already mentioned SPS WTO Agreement), that it is for each party to decide the applicable level of protection.

The formulation of the PP given by the Commission is consistent with that of the EU Court in implicitly assuming that the PP applicability is subject to a "plausibility test" for the concerns motivating a precautionary measure. The exclusion by the Court of legitimacy of a purely hypothetical approach to risk coincides with the requirement by the Commission of "reasonable grounds for concern" based on an objective preliminary scientific evaluation.

However, decision makers are faced with very diverse situations characterised by uncertainty or inadequate scientific evidence. Rigid definitions and rules for the PP would not be helpful. Each situation needs to be examined on a case by case basis in light of the duty placed by the EU Treaties on the institutions to ensure that measures falling within the policy areas under EU competence guarantee a high level of health and environmental protection.

The Commission in its communication has also set out guidelines for the application of the PP, when the conditions for its use are met:

- The measures envisaged must be proportional to the objective to achieve the chosen level of protection. This implies the choice of the least restrictive solution capable to achieve the objective;
- The measures should be non-discriminatory. Comparable situations should be treated in the same way;
- Measures should be consistent with the measures adopted in similar circumstances;
- The usual practices of risk management should be applied, including the evaluation of costs and benefits of all relevant options;
- The relevant scientific developments should be monitored and examined in order to revise or reconsider the measure if and when a more complete assessment becomes feasible.

Because of the very nature of the PP, measures taken under this principle should be considered and treated as temporary. Nevertheless, it should be observed that the nature of the issues addressed, of the scientific uncertainties in presence, and the size and severity of the potential impacts may sometimes require far-reaching initiatives which would not be easy to reverse. In such cases, priority should be given where possible to *no-regret measures* capable, on balance, of bringing benefits irrespective of the scientific confirmation of the risk in question.

The Precautionary Principle and International Agreements The PP has been controversial on the international level particularly in relation to trade disputes. The legitimacy of measures that are intended to protect health and the environment is internationally accepted under certain conditions, even if such measures may create trade barriers. Moreover, it is recognised that it is up to each country, and therefore to the EU, to set up its desired level of protection. Nevertheless, there are different views on the conditions for action in case of uncertainty or inadequate scientific evidence. In spite of the fact that the PP is mentioned, with various formulations, in several international agreements, there is still some disagreement across jurisdictions and scholars on its applicability and interpretation as a general principle of international law.

Examples of international agreements referring to measures taken in case of uncertainty are:

- The already mentioned SPS WTO agreement states that:

> in case where the relevant scientific evidence is insufficient, a Member may provisionally adopt sanitary or phytosanitary measures on the basis of the available scientific information, including that from the relevant international organisations as well as from sanitary or phytosanitary measures applied by other Members. In such circumstances, Members shall seek to obtain the additional information necessary for a more objective assessment of risk and review the sanitary or phytosanitary measure accordingly within a reasonable period of time.

Although this provision is close to the EU interpretation of the PP, its interpretation and application by the parties to the SPS Agreement remains subject to judicial review by an international Panel which pronounces on the substance of the risk issue based on its own views on precaution, in the absence of agreed criteria and standards.

- The Protocol on Biosafety of the Convention on Biological Safety (UN 2000) states that:

> lack of scientific certainty due to insufficient scientific information and knowledge regarding the extent of the potential adverse effects of a living modified organisms on the conservation and sustainable use of biological diversity in the Party of import, taking also into account risks to human health, shall not prevent that Party from taking a decision, as appropriate, with regard to the import of living modified organism in question as referred to in paragraph 3 above, in order to avoid or minimize such potential adverse effects.

- The so-called Rio Declaration adopted at the UN Conference on Environment and Development in 1992 (UN 1992) states that:

> In order to protect the environment, the precautionary approach should be widely applied by States according to their capability. Where there are threats of serious or irreversible damage, lack of full scientific certainty shall not be used as a reason for postponing cost-effective measures to prevent environmental degradation.

The International Panel on Climate Change-IPCC noted in its 2009 report (IPCC 2009):

> While the precautionary principle appears in many other international treaties, from a scientific perspective the concept of precaution is subject to a plurality of interpretations. To frame the discussions on precaution, three key points should be considered first.
>
> First, 'precaution' relates to decision-making in situations of deep uncertainty. It applies in the absence of sufficient data or conclusive or precise probabilistic descriptions of the risks …or in circumstances where the possibility of unforeseen contingencies or the possibility of irreversibility …is suspected.
>
> Second, in addition to that uncertainty/risk dimension, there is also a time dimension of precaution: the precautionary principle recognizes that policy action should not always wait for scientific certainty …
>
> Third, the precautionary principle cuts both ways because in many cases,… environmental choices are trade-offs between one risk and another risk. For example, mitigating climate change may involve more extensive use of nuclear power.

The Controversy on the Precautionary Principle Thousands of pages have been written on the Precautionary Principle, many of them to criticise it. It is not possible, and it would not be useful in this book to present a comprehensive analysis of the many criticisms of the PP. Nevertheless, some comments and clarifications about the most frequent criticisms may help to better understand what the PP is and is not.

Some underlying substantive reasons for the controversy on the PP have already been mentioned. In short, the PP:

- makes it more difficult to play delaying strategies by claiming that a risk is not demonstrated by a full risk assessment;
- creates pressure on risk promoters to invest in exploring risks and producing robust data at an early stage as part of the development work and to refrain from pursuing developments affected by significant lack of knowledge on their potential health and environmental risks;
- reverses the burden of proof in cases where developers are required on precautionary considerations to provide health and environmental data in order to obtain a pre-marketing authorisation;
- provides, in case of uncertainty, a legal basis for measures giving precedence to health and environmental protection in the public interest over sectorial economic interests.

The PP has been incorporated into the legal order of several jurisdictions, in different ways and with different formulations. For instance, the Principle is included in the French "Charte de l'environnement", the legal instrument framing the environmental regulatory policy in France. It is also a guiding principle for the environmental and health legislation in Canada, Australia and the UK. Nevertheless, many of the published criticisms directly or indirectly refer to the EU definition and practice. That is not surprising given the size of the EU market and the share of the EU in international trade. Behind the criticisms and their supporting arguments developed by several scholars, researchers or bodies, there are sometimes economic and trade interests or political objectives. Since the main arguments against the PP are quite repetitive, they are here reduced to their bare essentials, and summarised without referring to individual publications.

Some commentators claim that there is no such thing as a Precautionary Principle because dozens of different formulations can be identified in different texts. The existence of so many different formulations would in their opinion mean that the PP is not applicable. This is a very weak argument. In fact, the Precautionary Principle exists and is applicable as a principle of law where and how…it is established by the law as it is the case in the European Union as well as in various national jurisdictions. It is applicable in the EU in accordance with the provisions of the Treaties, the official documents clarifying its meaning and conditions of application and the doctrine of the Court. Provisions for precautionary action also exist under various international Conventions and Agreements and apply in accordance to their specific conditions and terms. There is no overarching world-level uniform formulation of the PP, as it is also the case for many other principles. That does not imply that the

PP does not exist or may not be effective and consistently applied in the jurisdictions where it is formally established. If the debate on the PP were less polluted by vested interests, it could be possible to reach greater convergence on the status and interpretation of the PP internationally.

It is also claimed that the EU Precautionary Principle is of no use to guide decision making because it is not defined in the Treaty. Other principles of law are not defined in significantly greater detail in the Treaties, and they are not considered to be useless or ineffective for that reason. A rigid definition for the PP would probably be counterproductive with respect to the objective of providing a basis for action in the variety of diverse situations characterised by uncertainty. Nevertheless, the guidelines adopted by the Commission and the pronouncements of the Court do provide clarity on the conditions under which the PP is applicable and on how it should be applied. A definition in the Treaty would be necessarily short and generic, not adding much to the established doctrine.

Certain authoritative scholars have claimed that the Precautionary Principle is intrinsically incoherent as its strict application would at the same time impose and prohibit any precautionary measure. In their view, that is because any alternative to the uncertain risk against which the measure is taken, would inevitably involve other uncertain ancillary risks that might not be accepted under the PP. Therefore, the PP would both require and forbid the precautionary measure. This argument refers to what certain authors call the "strong" formulation of the PP that would impose prohibiting a technology, product or development, based on hypothetical risks irrespective of the consequences, costs and adverse health and environmental effects resulting from the measure itself. At the same time the argument implicitly puts on the same level the uncertain risks dealt with by the PP and any purely hypothetical ancillary risk attributable to the consequences of the measure taken, even those based on mere conjecture. This argument is in fact based on an extreme interpretation of the PP which is not found in any jurisdiction, and certainly not in the EU. Under the EU approach, measures taken under the PP must comply with all the relevant principles, and all the options and alternatives for action must be evaluated in light of their effectiveness, consequences and costs. The criticism in question is based on extreme assumptions on the meaning of the PP and is therefore quite irrelevant in relation to the EU principles for the application of the PP to regulatory practice. In addition the assumption that any measure inevitably involves potential negative ancillary consequences that would fall under the criteria for the applicability of the PP, as defined by the Commission, is not convincing. The ancillary effects of a protection measure depend on the nature of the problem, the measure, the circumstances, but even more the action taken to prevent negative consequences. An example which is often given is the decision by certain countries to renounce to nuclear power plants, which is assumed to inevitably cause higher CO2 emissions from fossil fuels. Obviously, that consequence is in fact not inevitable on balance if nuclear power is gradually replaced by alternative non-carbon sources with overall lower environmental impacts. The possible overall risks associated with such technologies may well result to be much lower on balance than the risks displaced by the measures considered. Generalisation is not justified. A pragmatic, case-by-case assessment is needed.

It has been claimed that the adoption of the precautionary principle by the EU shows that Europe is more risk averse than its competitors, notably the US, with the consequence of a loss of relative competitiveness. In fact, while there are sectorial differences in the approach to risk, there is no solid evidence of a general, higher risk aversion in Europe. Comparative studies have shown that there are risk issues on which the EU is more risk-averse than the US and vice-versa cases where preventive policy and regulation are more stringent in the US (Wiener et al. 2011).

Finally, it should be noted that the controversy on the PP does in fact focus on a very limited number of well-known but complex cases in which a variety of considerations other than risk play an important part, notably GMOs, hormones in beef, endocrine disruptors. Drawing general conclusions on the principle itself from such cases and from some examples of clear misuse of it does not help progress toward a shared understanding and better application of the principle. We say more on the subject in our discussion on the role of science and evidence in regulatory policy, in Chap. 7.

5.6.2 *The Subsidiarity and Proportionality Principles*

Subsidiarity The requirement to ensure a high level of health and environmental protection applies across all the EU policies. Therefore, the EU has wide responsibilities for health and environmental risks. Nevertheless, not all risks are to be managed at EU level. Apart from areas that fall within the exclusive EU competence, risk policy and the related measures are subject to the subsidiarity principle, as are in fact all acts of the EU. Article 5 of the TEU states that:

> Under the principle of subsidiarity, in areas which do not fall within its exclusive competence, the Union shall act only if and insofar as the objectives of the proposed action cannot be sufficiently achieved by the Member States, either at central level or at regional and local level, but can rather, by reason of the scale or effects of the proposed action, be better achieved at Union level. The institutions of the Union shall apply the principle of subsidiarity as laid down in the Protocol on the application of the principles of subsidiarity and proportionality. National Parliaments ensure compliance with the principle of subsidiarity in accordance with the procedure set out in that Protocol.

In the areas of Union exclusive competence, where the subsidiarity principle does not apply, only the Union may legislate and adopt legally binding acts. Within the exclusive competence category, the conservation of marine biological resources under the common fisheries policy and the common commercial policy are of particular relevance for the EU risk policy. In particular, under the common commercial policy, the EU has the exclusive competence to conclude and manage trade agreements that play a growing part in setting the criteria and conditions for trade-restrictive measures.

The Treaties confer on the Union a wide range of competences shared with the Member States. In such areas, as we have already explained, both the Union and the Member States may legislate and adopt legally binding acts, but the Member States shall exercise their competence only to the extent that the Union has not exercised its competence. There are several areas of shared competence that are relevant for risk policy, notably the internal market, the social policy, agriculture and fisheries, the environment, consumer protection, transport, energy and the common safety concerns in public health matters.

Finally, we have seen that in certain areas, and under the specific conditions laid down in the Treaties, the Union has competence "to carry out actions to support, coordinate or supplement the actions of the Member States, without thereby superseding their competence in these areas". Here the most relevant areas for risk policy are protection and improvement of human health, industry and civil protection.

The application of the subsidiarity principle is subject to the provisions set out in a protocol that includes both procedural and substantive requirements. In particular, the Protocol states that:

> Draft legislative acts shall be justified with regard to the principles of subsidiarity and proportionality. Any draft legislative act should contain a detailed statement making it possible to appraise compliance with the principles of subsidiarity and proportionality. This statement should contain some assessment of the proposal's financial impact and, in the case of a directive, of its implications for the rules to be put in place by Member States, including, where necessary, the regional legislation. The reasons for concluding that a Union objective can be better achieved at Union level shall be substantiated by qualitative and, wherever possible, quantitative indicators. Draft legislative acts shall take account of the need for any burden, whether financial or administrative, falling upon the Union, national governments, regional or local authorities, economic operators and citizens, to be minimised and commensurate with the objective to be achieved.

Proportionality The proportionality principle applies in all EU policy areas and to all types of regulatory measures. It applies to all EU risk regulatory measures. The Treaties, Art.5(2) TEU, state in particular:

> Under the principle of proportionality, the content and form of Union action shall not exceed what is necessary to achieve the objectives of the Treaties.

This principle implies two conditions:

1. The measure is necessary in order to achieve the objectives of the Treaties. Normally this condition is met in the case of measures which are:
   - in conformity with the conferral and subsidiarity principles (they fall under EU competence; action at EU level is justified and the type of measure or initiative is conform to what the Treaties prescribe for the relevant category of competence);
   - aimed at ensuring a high level of protection, as prescribed by the Treaties, by preventing or reducing health, safety or environmental risks;
   - limited, as for their content, to set out the initiatives or provisions strictly needed to ensure the effective achievement of the intended objectives.
2. The form of the action does not "exceed" what is necessary to achieve the objective. That means that the instrument chosen should be the least restrictive and most efficient one, compatible with the effectiveness of the measure, the applicable principles and the provisions of the specific legal framework. Therefore, all range of instruments and tools should be considered, including in particular product labelling, consumer information schemes, voluntary agreements, standards etc.

5.7 Assessing Risk Management Options

Assessing options for dealing with risks is at the core of the risk management function. We have seen that the EU Better Regulation Guidelines provide the methodological and procedural framework for risk management. There are however some aspects that are very specific to the assessment of risk management options. Risk-reduction benefits of the measures considered must be weighed against the lost benefits of the products or technologies which are subject to possible restrictions. Since the measures considered may themselves cause ancillary risks, risk-risk comparison is often needed. It may be impossible, not practical or disproportionately costly to eliminate completely all risks: risk managers may need to assess options against acceptability criteria for the residual risk.

There is no rigidly defined methodology for the risk management analytical function. We wish to introduce here some of the main risk management issues.

5.7.1 Risk-Risk Comparison

We have already mentioned the issue of risk-risk balance in Chap. 4 on risk assessment. Measures intended to prevent or reduce certain specific risks may, as unintended side-effect, create a different risk. Or they may bring additional ancillary benefits. We have discussed the case of phthalates in PVC medical devices. In that case, the conclusion was that a measure restricting the use of phthalates in order to eliminate the health risks associated with their release might, in the absence of safe and equally effective substitutes, create a different health risk by negatively interfering with certain life-saving medical practices. There are many concrete examples of risk management measures that may cause what are often called "ancillary risks", or, in the opposite direction, may bring additional risk reduction benefits beyond the specific objective of the measure considered: air pollutant emission standards for industrial plants may increase the risk of soil pollution if the waste resulting from flue gas treatment are not properly conditioned and disposed of; conversely, measures intended to cap CO_2 emissions for mitigating climate change risks may also contribute to reduce pollution and accident risks related to the extraction, transport and combustion of fossil fuels.

In fact, there are various categories of possible ancillary risks and benefits. When the use of certain products, substance or technologies is prohibited or restricted, their substitutes may cause unintended ancillary risks, if such risks are not in turn properly regulated. The ancillary risks may be of the same or of a different nature of the risk displaced, and may fall on the same or a different group of population. For example, in the case of phthalates in medical devices, the issue is one of the health-health balance, for the same exposed group. Similarly, renouncing to add chlorine to drinking water may reduce the risk of cancer, but increase microbiological contamination risks. When fuel efficiency standards for cars were discussed in the US, there were concerns that, while reducing atmospheric pollution and its related health impacts, they might cause safety risks by reducing the passive safety of cars, if lighter vehicles were introduced as a consequence of the new standards. CO_2 reduction measures for mitigating climate change may induce a reduction of coal extraction with a decrease of the related work accidents, an example of ancillary benefits.

Ancillary risks may be a direct consequence of the measure taken, like in the example of chlorine in drinking water, or may indirectly derive from it, for example because of the induced substitution of a banned substance with one posing different, perhaps less studied risks. They could also depend on the so-called "lulling effect", a careless behaviour caused by the false sense of safety induced by the measure, for example the lack of surveillance of babies by parents due to the introduction of child-resistant packaging of dangerous products. Finally, some authors have highlighted the need for assessing a very indirect health-wealth-health balance in order to take into account the possible negative effects on health of measures addressing health risks but at the same time reducing wealth and therefore putting at risk the health conditions of the less wealthy. We say more on this further on.

A variety of different combinations are possible for the risk-risk balance, involving health, environmental or safety risks. As a consequence, it may be difficult to establish an overall balance with a single metric. For instance, use of electric cars may improve air quality on the local level, but increase the risk of climate change if the electricity is produced burning coal. In the risk management process, it is important to examine the main side effects, in particular the possible unintended risks, which may derive from the implementation of the options considered.

The first step of a risk-risk comparison is the identification of all consequences of the intended measure. The identification of ancillary risks needs building scenarios for the way in which the measure may be expected to be received and applied by the relevant actors. The examination of the potential consequences of policy and regulatory options helps deciding the appropriate scope of the measures envisaged. For instance, if there are reasons to expect that a ban on a particular substance used as refrigerating fluid, contributing to stratospheric ozone depletion, might lead to the introduction of substitutes with similar depletion properties, the measure should ensure that all substances for the application in question be tested and only allowed if proven safe in that respect.

In the example of ancillary risks mentioned above, a condition ("if…") has been mentioned. In fact a potential ancillary risk does not necessarily need to materialise. Much depends on the specific circumstances. For example, air pollution standards for industrial installations do not need to lead necessarily to increased soil pollution from the disposal of flue-gas treatment waste, if a proper regulatory framework for waste management is in place and is effectively applied.

The issue of risk-risk balance should not be dealt with on a purely hypothetical approach to ancillary risks, based on speculations on unsubstantiated possible negative consequences of the measures envisaged. Applying the precautionary principle to purely hypothetical ancillary risks would be incompatible with the EU guidelines for the application of the PP. It is not reasonable to assume that risk reduction measures would cause in all circumstances significant ancillary risks. Speculation on hypothetical ancillary risks should not be taken as a justification for not acting against well identified and concrete risks.

As illustrated in Chap. 3, the EU has established since many years a comprehensive policy and regulatory approach to environmental, health and safety risks. That includes in particular the integrated pollution control legislation, comprehensive regulatory frameworks for chemicals, food and wastes, general product safety requirements for non-food products, in addition to a wide body of issue-specific legislation and standards. Moreover, differently from those jurisdictions where certain agencies with narrow sectorial competences might possibly act without regard to the possible negative effects outside their domains, in the absence of adequate coordination, EU-level measures are taken through a decision-making process involving a single line of institutions facilitating a comprehensive and coordinated approach. The instruments are therefore in place in the EU in order to facilitate an integrated approach to risk-risk issues and prevent unexpected ancillary risks, while maximising the win-win solutions bringing, where possible, additional health, safety or environmental benefits.

It should be noted that the risk-risk issue has been a controversial subject, notably in the US. The claim of scarce consideration of ancillary risks by sectorial regulatory agencies has been an argument for invoking the establishment of formal requirements for integrated risk assessment. A debate has developed on the justification and feasibility of such a request. Certain stakeholders and commentators have seen it as an attempt to put a brake on the regulatory action of agencies like the US-EPA or the US-FDA. The concern was that far-reaching requirements for the identification and assessment of any (even highly hypothetical) ancillary risk would impose an unreasonable burden and expose the intended measures to endless challenges and re-examination. As said above, consideration of the balance between risks is in fact important and necessary when there are significant ancillary risks with a well identifiable causal link with the measure envisaged. Normally such ancillary risks can be identified with a relatively simple screening. Attempts to assess very indirect and hypothetical ancillary risks would not be practical and cost-efficient. Certain scholars have suggested that any risk prevention measure by absorbing resources has a negative impact on wealth, and therefore indirectly a negative impact on the health of certain groups of the population (reversing the slogan "health is wealth" into "wealth is health"!). They claim that such possible negative impact should be brought into the risk-risk balance, submitting the measures considered to the outcome of complex and somewhat obscure economic models. The rationale and practicability of such an assessment is arguable and introducing a requirement in that respect would be difficult to accept by democratic institutions confronted with growing requests of transparency on the reasoning behind their decisions. In a scientific, fact-based approach, this kind of academic theories have a suspect ideological flavour and do not seem suitable for supporting a pragmatic and result-oriented assessment. Moreover, there are many examples of risk prevention measures that may instead have a positive effect on wealth. For instance, air quality standards that may, as indirect effect, stimulate the replacing imported fossil fuels with domestically produced renewable energies can, depending on the market response, increase on balance wealth, even in the absence of net cost-savings, by creating opportunities for business and employment in the countries where they apply.

As already mentioned, one of the main difficulties in establishing a risk-risk balance is that of the metric to be used when the balance is between risks of different nature. How to compare the reduced risks of climate change obtained by a switch to nuclear power with the increased radiation-exposure risks associated with potential accidents in the nuclear industry? On the one hand we have a variety of far-reaching, long term and global health and non-health impacts, whereas, on the other hand, nuclear is overwhelmingly characterised by more localised health and safety risks. A complex, multidimensional comparison is necessary in case of incommensurable risks. However, there are simpler cases, notably when both target and ancillary risk are of the same nature, notably when in both cases the risk may be expressed as an additional death rate.

5.7.2 *Risk-Benefit Evaluation*

When examining the options for action, risk managers may need to establish the balance between the risks and the benefits of a product, technology or activity. Typical examples are provided by measures authorizing or restricting medicines, medical devices or medical practices, which may be assessed under both the risk-risk and risk-benefit angles. For example, a pharmaceutical product may expose patients to risks of negative side-effects, while providing the therapeutic benefit against the disease for which it was developed and tested. Similarly, medical devices and surgical interventions may involve risks that need to be weighed against their expected benefits. The balance between risks and benefits is often difficult to establish. Both risks and benefits may be affected by uncertainties. It is particularly difficult to compare quantitatively risk and benefits which are incommensurable and cannot be expressed using a common metric. Nevertheless, we all make such comparison qualitatively and intuitively, for instance accepting the risk of accident when travelling by car in view of the benefit of speed and convenience. Multi-criteria techniques are available aimed at facilitating the establishment of comprehensive risk-benefit balances in such cases.

The issue of risk-benefit balance is particularly relevant, and systematically considered, in decisions concerning pharmaceutical products. In the EU, the European Medicine Agency (EMA) is given the task of examining medicines proposed for authorisation for their efficacy, safety and quality, notably in order to establish whether any side-risk is justified by the health benefits expected. Medical research and clinical trials are also subject to authorisation based on consideration of their risks and potential benefits. In general, authorizations are subject to the risks to participants being minimized by appropriate research design and by procedures which do not unnecessarily expose subjects to risk. The overarching condition is that risks to subjects are reasonable in relation to anticipated benefits and the importance of the knowledge improvement that may reasonably be expected. The risks of medical research may be physical, psychological, social, and economic. Inevitably, qualitative considerations intervene in the assessment of risks. For instance the definition of minimal risk given by the US Department of Health & Human Services (Code of Federal Regulations, Title 45 Public Welfare, Department of Health and Human Services, Part 46 Protection of human Subjects, revised 2009):

> Where the probability and magnitude of harm or discomfort anticipated in the proposed research are not greater, in and of themselves, than those ordinarily encountered in daily lives of the general population or during the performance of routine physical or psychological examinations or tests.

In certain regulatory areas, consideration of the limited benefits at stake leads to a conservative approach to risk. For instance, the EU legislation on cosmetic products has introduced a strict precautionary approach in consideration of the nature and function of the products regulated.

Risk-benefit assessment is also relevant for risk management of food. Typical examples are the assessment of the balance between the health benefits of a diet including consumption of fish and the potential risks due to contamination of certain species of fish in certain regions by PCBs, dioxins or mercury, or the balance between health benefits and toxicity risks of the addition of vitamins and micronutrients to food.

Finally, when considering risks and benefits, managers often need to factor in uncertainties on both sides. It may be very difficult to evaluate benefits at stake in a deterministic way. How to estimate the possible lost benefits due to a stringent regulation imposed on the development of a new technology, say synthetic biology? Any calculation would necessarily require highly hypothetical assumptions on the future developments of the technology, the way in which developers might respond to the regulatory requirements, the future availability of alternatives etc.

In this section we have used the expression "risk-benefit balance" and in the next one we will talk of risk-cost balance. It would perhaps be more appropriate to talk of the relationship between the value of risk reduction and the cost of the measure. That would avoid the wrong impression that any such balance could be reduced to a single figure of "net" cost or benefit. Clearly, risk-benefit methods are nevertheless very relevant for evaluating and comparing the loss related to the potential adverse outcomes associated with a risk (what is at stake), the value of a risk reduction measure (what can be gained or saved) and the cost of the measure. There is no shortage of guidance on cost-benefit assessment. In particular, the Impact Assessment Guidelines of the European Commission provide tools for such evaluations. Nevertheless, monetary cost-benefit methods are not sufficient for comprehensively assessing all the relevant aspects of the societal losses related to the potential adverse impacts of risk and the value of the risk-reduction measures. A discussion of the potential and limits of cost-benefit assessment is not within the scope of this book. It suffices to say that for risk-benefit and risk-cost assessment a multi-criteria approach, involving consideration of both monetary and non-monetary aspects is needed in order to provide decision makers with the complete picture in a comprehensive and structured manner.

5.7.3 Risk-Cost Evaluation and Its Limits

Risk-reduction measures involve direct and indirect costs, which are normally expressed in monetary terms. In turn, they bring the benefits attached to the reduction of risks, which is the main objective of the measure. Depending on the nature of the risks, certain aspects of those benefits may be expressed, more or less easily and completely, in monetary terms. For instance, a monetary value may be relatively easily attached to the reduction of property-damage risk. Valuing the reduction of health risks or of the risk of death is much more difficult and controversial. Obviously there are cost-aspects related to health-risk reduction, for instance the healthcare savings or the reduction of lost working days or of lost income, but health is considered a good and a right on its own, which cannot be exhaustively represented through its monetary attributes. Therefore, in order to relate the monetary cost-aspects to the risk-reduction benefits (how much "good" is achieved for each euro or dollar spent), specific non-monetary indicators, notably "QALYS"

(Quality-Adjusted Life Years) and DALY (Disability-Adjusted Life Year), have been established and are used in the risk management assessment of options.

Quality-Adjusted Life Year (QALYS) QALYS is a measure of the state of health of a person or group in which the benefits, in terms of length of life, are adjusted to reflect the quality of life. One QALY is equal to 1 year of life in perfect health. QALYs are calculated by estimating the years of life remaining for a patient following a particular treatment or intervention and weighting each year with a quality of life score (on a zero to 1 scale). It is often measured in terms of the person's ability to perform the activities of daily life, freedom from pain and mental disturbance (Source: https://www.nice.org.uk/glossary).

Disability-Adjusted Life Year (DALY) One DALY can be thought of as one lost year of "healthy" life. The sum of these DALYs across the population, or the burden of disease, can be thought of as a measurement of the gap between current health status and an ideal health situation where the entire population lives to an advanced age, free of disease and disability. DALYs for a disease or health condition are calculated as the sum of the Years of Life Lost (YLL) due to premature mortality in the population and the Years Lost due to Disability (YLD) for people living with the health condition or its consequences (Source: WHO).

Healthy Life Years (HLY) Similar to QALY, HLY measures the number of quality adjusted remaining life years per person.

QALYS, DALY and HLY have been used in different contexts, notably for the assessment of health technologies and are not without problems due to the complexity and uncertainty often affecting their calculation and the doubts about the underlying assumptions and weighing criteria. For example, the "ECHOUTCOME" project financed by the European Commission (ECHOUTCOME 2013) based on a survey of 1,361 subjects in the UK, Belgium, France and Italy, found some invalid theoretical assumptions underlying QALYs, concluded that divergent QALY results can be generated using the same dataset and advised against using QALYs for health decisions.

Nevertheless, the use of indicators like QALYS and DALY for the assessment offer a way to express quantitatively the benefits of a health risk reduction measure and to compare different options by use of the same metrics. This is why these indicators are among the non-monetary indicators included in the Impact Assessment Guidelines (2009 and 2015) of the European Commission. In practice, health benefits are often assessed in a qualitative way. The EU approach is to try to quantify both the monetary and non-monetary aspects of health measures. Sometimes, the best option can be identified on the basis of a cost-monetary benefits balance, with non-monetary indicators as additional information. However, the EU approach uses monetary expression of health damage or benefits as a technical means to compare options, without aiming at attributing an absolute monetary value to life or to a given health status. Even when a monetary value is attributed to life for comparative purposes, that is only done on a statistical basis and never in relation to individuals.

5.7.4 Hazards Versus Risks in Risk Management

The EU policy-approach in the areas of health, safety and environmental protection is in general based on consideration of "risk" as defined in Chap. 2. We have already discussed the importance of the distinction between risks and hazards. Nevertheless, we have noted that action may need to be taken in light of hazards alone for certain particular objectives or in certain circumstances. For instance, the EU legislation on classification and labelling of dangerous chemical substances is a regulatory measure based on consideration of hazards. Its objective is to ensure that the intrinsic hazardous properties of chemicals be identified, studied, classified and communicated. Classification and labelling is the basis for preventive measures, where appropriate, depending on the applications of the chemicals considered and the resulting exposure, and for the safe handling, transport, use and storage of the substances in question. Additionally, under the REACH Regulation substances with certain hazardous characteristics (substances of high concern) may be placed on the market and used only in authorized applications. Here, the precautionary principle is applied in order to reverse the "burden of proof". When requesting the authorization for a specific application of one of those substances, it is for the risk-promoter to demonstrate that the resulting risk is acceptable and justified in light of the expected benefits and account taken of possible alternatives.

Bans or restrictions must be introduced based on hazard considerations alone when it is impossible, impractical or disproportionately costly to ensure, with a sufficient degree of confidence, the exposure restrictions under which an unacceptable risk could be prevented. That is in particular the case of applications of certain highly dangerous substances for which it would be very difficult in practice to enforce compliance with safe exposure limits. Measures of this type are sometimes criticised as a breach to the risk-based approach. It is in fact the impossibility to control the risk that justifies acting on hazards in such cases.

A notable example is provided by the controversial EU legislation on pesticides, in its formulation of 2009 (EU 2009). Certain provisions of the relevant Regulation are driven by the intrinsic properties of the substances considered. For instance, the use in pesticides of certain carcinogenic, mutagenic, repro-toxic or persistent substances is banned or is only allowed if the resulting exposure is "negligible". When these restrictions were adopted following strong political pressure by the European Parliament, industry reacted negatively. For example, Julian Little, public and government affairs manager for Bayer CropScience and Chair of communications for the UK Crop Protection Association, claimed in a public statement of January 2009[1] that the regulatory approach was flawed because:

---

[1] Reported in the website of the UK Royal Society of Chemistry http://www.rsc.org/chemistryworld/News/2009/January/23010902.asp.

> Pesticides are by their nature biologically active, and therefore if you go out looking for hazards associated with them, chances are you'll find one. By ignoring such things as safe working practices and even the dosage involved, the new rules will disregard products which have been proven safe over years of use and by a regulatory system which is second to none in the world.

However, the point of the EU legislator was that it did not make much sense to rely upon exposure limit values for products intended for being spread in the environment by a multitude of operators, with uncertain fate, under conditions that are practically impossible to control. In these particular cases, the legislator concluded that it was incompatible with the Precautionary Principle, in light of both the experience[2] and a-priori considerations, to allow the use of substances of such high concern in applications involving a wide dispersion into the environment, relying upon a risk-based approach affected by a great number of uncertain assumptions, hypotheses and extrapolations on the fate of and exposure to these substances.

Significantly, the EU Court of First Instance had annulled in 2007 a Commission Directive including "paraquat" (an active substance which was a component of one of the three most widely-used weed-killers in the world) in the list of substances allowed for use in plant protection products, considering that the Commission had failed to demonstrate that such highly toxic substance would have been used in a way consistently limiting the exposure below the safe level and concluding:

> The Commission's handling of the file does not satisfy the applicable procedural requirements and the directive fails to satisfy the requirement of protection of human and animal health (European Court of Justice 2007).

5.7.5 *Dealing with Uncertainty and Variability in Risk Management*

In this book, we frequently mention the importance of uncertainty in risk analysis. It is clear from what we have discussed so far that risk management most often takes place under a significant degree of uncertainty. We wish to remind here the distinction between the uncertainty that is intrinsic in the very concept of risk and the uncertainty on the existence of a risk, its probability to materialise, its size, distribution and timing. The first type of uncertainty may be approached in a quasi-deterministic way. For instance, if I know that the probability to lose 10.000 euros in a repetitive series of financial operations is 0.3 (30 %), I can calculate the risk of loss as $0.3 \times 10.000 = 3.000$ euros and compare this figure with the cost of an insurance product. However, for the second type of uncertainty, risk managers are

[2] For instance, in the "paraquat" case referred to below, the Court found that the assumptions made by the Commission on the maximum level of exposure were contradicted by a study of real cases (see also below).

provided only with qualitative statements or with estimates or distributions of probabilities often based on expert judgement.

Moreover, when the science is weak or controversial and the uncertainties are high, a third layer of uncertainty is added, the uncertainty on the robustness of the uncertainty assessment itself. In other terms, risk managers are left with doubts on the level of confidence that they can place on the expert assessment. Complex mathematical methods and models have been developed to aid decision in presence of uncertainties. Quantitative theories of decision are well developed. Nevertheless, they do not play in practice a significant role in the political decision making process, notably because the use of decision models adds a further layer of complexity and opacity to the underlying risk issues.

5.7.6 The Concept of Acceptable Risk

Let's now add to what already said in this book on risk acceptability. When examining the options for risk-prevention or risk-reduction measures, risk managers and decision makers are often faced with the need to decide how much safe is safe, taking into account the effects of the various options on the target risk, the appearance of ancillary risks, the residual risks that cannot be eliminated, or that can only be eliminated at very high costs or with serious negative side-effects.

Often, the measures taken leave some residual risk. In other cases, no measure is taken considering that the risk is "acceptable". In such cases, the question is what is an acceptable risk, under the specific circumstances in presence. We need to make here some considerations on the issue of acceptability, in addition to what we have said on the subject in Chap. 2.

Let's start by dispelling the confusion around the "impossibility" of zero risk. It is a well-established mantra that "zero-risk-does-not-exist". However, there is no justification for the claim that any risk prevention measure necessarily leaves some residual risk. Whether a complete elimination of a specific risk can be achieved depends on the nature of the risk and on the available options. For instance, it is sometimes possible to replace a toxic component of a product with a well-studied, demonstrated non-toxic one, with similar functional properties, thus reducing the specific toxicity risk to zero without any appreciable adverse side-effect. Often the "no zero risk" argument is used for protecting those interests which inevitably are affected by measures that involve recourse to a safe alternative and create winners and losers in the business arena, in what is perceived in the short term as a zero-sum game.

When the zero-option is not technically available, not practical or disproportionately costly, risk managers need to determine the acceptable risk-level. Risk acceptability is indeed at the heart of risk management. The concept of acceptable risk refers to the probability and size of human health, environmental damage or property loss that can be tolerated by an individual, household, group, organization, community, region, state, or nation. The objective is to set the optimal balance

Table 5.3 Factors affecting social acceptability of risk

Risks perceived as more acceptable	Risk perceived as less acceptable
Voluntarily incurred	Imposed by others
Subject to individual's control	Uncontrolled by the exposed people
Accompanied by benefits for the people exposed	Without significant, direct benefits
Distributed fairly	Unfairly distributed
Caused by natural events	Manmade
Characterised by a limited impact	Catastrophic
Controlled by a trusted organisation	Organisation at the source of risk untrusted
Familiar or already experienced	New or unknown
Affecting people able to manage it	Affecting people unable to manage it
Directly relevant for us	Not likely to directly affect us

between the level of residual risk and the need to ensure the functional requirements, benefits and cost effectiveness of the product, technology, project or activity considered. In analyzing the concept of acceptable risk, one has to take into account the multidimensional nature of risk, as explained in Chap. 2. Political decision makers are accountable to the public. In deciding what is acceptable they need to consider for whom the risk is or is not acceptable and how a specific risk is perceived and weighed by those affected.

Table 5.3 complements the one that we have presented in Chap. 2 and summarizes again some of the main factors determining the social acceptability of risk:

The table shows that risk acceptance is a highly situation-dependent concept.

There are many ways to set acceptable risks, which depend on the nature of the risk, the organization setting the acceptance thresholds and its perspectives and objectives. For instance, organizations setting acceptable levels of occupational risk often consider, in addition to the likelihood and frequency of adverse events or accidents, indicators like the seriousness of worker injuries (for instance injuries requiring intervention beyond first aid), the number of lost working days, the duration and cost of loss of facility operational capability etc.). In general, the combination of frequency and gravity is considered and the borderline of acceptable risks may be represented in a purely indicative manner as shown in Table 5.4 where the acceptable risk region is in green, the unacceptable one in red, while the position on the yellow region is undetermined:

It is in practice difficult to express quantitatively the risk in terms of a single indicator (for instance the probability and number of additional deaths in a given exposed population) and even more to fix, under the EU institutional conditions, uncontroversial and legally defensible quantitative risk-acceptability thresholds. In some jurisdictions, ranges are set for the acceptability of certain risks. An example is given in Tables 5.4 and 5.5 which shows the ranges of excess cancer and non-cancer hazards (expressed as Hazard Quotient, HQ), determining acceptance, further assessment or immediate action, which were stablished under the US-CERCLA legislation (US 1980).

Table 5.4 Risk tolerability

RISK TOLERABILITY REGIONS						
PROBABILITY	Very likely					
	Likely					
	As likely as not					
	Unlikely					
	Very Unlikely					
		Negligible	Low	Moderate	High	Very High
		IMPACT				

Table 5.5 CERCLA risk grading

Action	No further study warranted	Conduct a site-specific risk assessment to determine if clean-up warranted	Response action warranted
Concentration	Zero concentration	Regional screening levels	Removal management level
Criteria:		10^{-6}	10^{-4}
Excess cancer		HQ = 1	HQ = 3
Hazard quotient			

In the case of cancer risk, the level of 10^{-6} (one additional cancer in one million) is often mentioned as the threshold for acceptability. This figure appeared in the US as an essentially zero level of risk to which a *de minimis* approach may be applied. Its origin is not very clear. There is no known rationale supporting the choice of this value, although sometimes it is claimed that 10^{-6} is the probability of being hit by lightning, not a very convincing explanation because such probability may vary depending on regions and lifestyles and in any case there is no particular reason to prefer as a reference threshold the risk of lightning to other low risks. The figure of 10^{-6} has not been endorsed in any legal text in the EU, but it is sometimes used as a reference in risk and impact assessment. Sometimes, the risk under consideration is quantitatively compared with risks related to other sectors, products or activities. For instance, the risk of death following a nuclear accident is compared with the risk of death in car accidents, assumed to be socially "accepted". This approach is flawed on various grounds. First, there is no reason to accept an additional risk just because it is no greater than an existing one. Risk includes probability in its very definition, and probabilities of independent events are additive. The public would not be compelled to accept an additional risk just by reassurance that it will be lower or comparable to the ones already present. Second, the comparison of risks across sectors based on a single metric fails to catch the wide differences that exist in risk perception, as we have mentioned before, depending on the specific nature and profile of the risk. Third, the risks taken as socially accepted are instead often conceived by those affected as de-facto imposed, given the circumstances applicable. Driving a car may be for many less a choice than a practical need for fulfilling work duties or other imperative obligations. Finally, often the comparison fails to factor-in the different distribution across the population of risks caused by different sources.

An important approach to residual risk is the so-called ALARA (As Low As Reasonably Achievable) principle or its ALARP (As Low As Reasonably Practical) variation. ALARA is a risk management concept that has mainly been developed and applied to radiation protection. It is defined by the US Nuclear Regulatory Commission (http://www.nrc.gov/reading-rm/doc-collections/cfr/part020/part020-1003.html) as:

> ALARA is an acronym for "as low as (is) reasonably achievable," which means making every reasonable effort to maintain exposures to ionizing radiation as far below the dose limits as practical, consistent with the purpose for which the licensed activity is undertaken, taking into account the state of technology, the economics of improvements in relation to state of technology, the economics of improvements in relation to benefits to the public health and safety, and other societal and socioeconomic considerations, and in relation to utilization of nuclear energy and licensed materials in the public interest.

The ALARA principle is also referred to in the EU legislation on protection against the dangers arising from exposure to ionizing radiation. The EU approach to protection against damage from ionizing radiation is based on the assumption that it is not possible to establish a threshold for biological damage caused by radiation and therefore there is no safe exposure level. The legislation therefore imposes three requirements on any exposure practice: justification, optimization and dose limitation which are summarized in Box 5.1.

Box 5.1

General Principles on Exposure Practices according to EU Directive 2013/59/EURATOM on protection against the dangers arising from exposure to ionizing radiation.

Justification: Decisions introducing a practice shall be justified in the sense that such decisions shall be taken with the intent to ensure that the individual or societal benefit resulting from the practice outweighs the health detriment that it may cause. Decisions introducing or altering an exposure pathway for existing and emergency exposure situations shall be justified in the sense that they should do more good than harm

Optimisation: Radiation protection of individuals subject to public or occupational exposure shall be optimised with the aim of keeping the magnitude of individual doses, the likelihood of exposure and the number of indi-

(continued)

Box 5.1 (continued)
viduals exposed ***as low as reasonably achievable*** taking into account the current state of technical knowledge and economic and societal factors. The optimisation of the protection of individuals subject to medical exposure shall apply to the magnitude of individual doses and be consistent with the medical purpose of the exposure. This principle shall be applied not only in terms of effective dose but also, where appropriate, in terms of equivalent doses, as a precautionary measure to allow for uncertainties as to health detriment below the threshold for tissue reactions.

Dose limitation: In planned exposure situations, the sum of doses to an individual shall not exceed the dose limits laid down for occupational exposure or public exposure. Dose limits shall not apply to medical exposure.

Level of Risk	ALARP *As Low As Reasonably Practicable*	
High Risk	Region of Unacceptable Risk	Risk cannot be justified on any ground
Medium Risk	Region of ALARP	Risk is tolerable if risk reduction not practicable or involving costs grossly disproportionate to the benefit achieved
		Risk is tolerable if cost of reduction would exceed the improvement gained
Low Risk	Region of Broadly Acceptable Risk	Risk is acceptable, no need for considering cost and benefits of further reduction measures

Fig. 5.1 The ALARP principle

The ALARP principle, "as low as reasonably practicable", puts the accent on the cost of further risk reduction as the main criterion to identify the "reasonable" level of acceptable risk. It is used in the fields of system safety and safety at the workplace, mainly in the UK. According to ALARP the residual risk shall be as low as reasonably practicable in the sense that the cost involved in a further reduction of the risk would be disproportionately high compared to the benefits achieved. The concept is illustrated by the scheme in Fig. 5.1.

Under the EU regulatory approach, the final judgement and decision on the acceptability of a residual risk is made by the EU political institutions in light of the legal, technical, economic, societal aspects of the case. In general, there is no predetermined quantitative acceptability threshold which could be mechanically applied. Calculations valuing life and health conditions or showing the position of a risk in relation to certain death or disease incidence may inform but do not determine the political decisions.

At the end of this brief discussion on the acceptability of risk, we would like to repeat here again the definition of safety given by Lowrance that we have mentioned in Chap. 2 “a thing is safe if its attendant risks are judged to be acceptable”. Safety and acceptability are matter for judgement. The legitimacy of any threshold of risk acceptability depends on the informed consent of those who bear the risk. Transparency and participation are the key words in that respect. Our discussion on governance, further on in the book, will highlight the procedural aspects of participation in risk decisions.

5.7.7 *No-Regret Measures and Provisional Measures*

This Chapter has shown the complexity of risk management issues. Tools are available to identify and assess evidence and provide a factual basis for risk management. Guidelines, principles and methods have been established to help identifying and comparatively assess options for their advantages and disadvantages. Nevertheless, as experience shows, there cases where science can offer only a limited support for the identification, assessment and selection of policy and regulatory options. It is sometimes possible in such cases to identify options for action that can be justified by their ancillary benefits. A typical example, already mentioned, is that of the climate-change mitigation measures that can also contribute to cost-saving energy efficiency improvements and to increased supply security. Those are no-regret measures that might stand the cost-benefit test even if the risk that they were intended to address resulted, in light of new scientific knowledge, to have been overestimated. Finally, in case of important uncertainties and when acting on the basis of the precautionary principle, risk managers and policy makers should consider pragmatic and adaptive regulatory approaches, including strategies and measures that could be reconsidered, modified or withdrawn, minimising costs while ensuring effectiveness. Often, the most efficient strategies are those that start with no-regret measures, invest in improving the knowledge basis, include review and revision mechanisms and proceed gradually to more substantial initiatives, or instead to revise and adapt the ones already taken, in light of the results achieved, the difficulties encountered and any new relevant data and knowledge.

Summary

1. Risk management is the process of identifying the relevant risk policy options and assessing them for their effectiveness and efficiency in relation to the required level of protection and the other relevant policy aims as well as the applicable principles and criteria, taking into account the information provided by risk assessment, the costs and benefits of the various options and the legal and societal aspects. It is a technical function aimed at informing and supporting the political decision making.
2. Risk management and risk assessment are two distinct functions of risk analysis. The integrity of risk assessment and its independence from any undue influence

or interference must be preserved. Nevertheless, a sustained dialogue between managers and assessors is necessary. Risk assessment should be framed within a shared risk assessment policy ensuring that all the value-laden aspects of risk assessment are addressed in a transparent manner and are subject to policy maker scrutiny.

3. The risk management function for the EU risk policy and regulatory measures must provide support to a complex, multilayer decision making process, articulated in different procedures depending on the applicable legal basis, involving the three EU institutions (Commission, Parliament and Council), with an approach open to the input from a variety of stakeholders and against a background of 28 Member States with diverse situations, priorities and interests. The EU risk management is framed within the structured approach of "Better Regulation" established by the Commission and shared with the other EU institutions.
4. Several principles established by the EU Treaties apply to EU risk policy and regulatory measures, notably subsidiarity, proportionality and precaution. Although the EU has wide responsibilities for health, safety and environmental protection, under the subsidiarity principle it acts only if and insofar as the objectives of the action cannot be sufficiently achieved by the Member States. Moreover, according to the proportionality principle the content and form of the Union action shall not exceed what is necessary to achieve the objectives of the Treaty. These are horizontal principles of the Treaty.
5. The Precautionary Principle is specifically relevant for measures addressing health, safety and environmental risks. According to guidelines established by the European Commission, it addresses those specific circumstances where scientific evidence is insufficient, inconclusive or uncertain and there are indications through preliminary objective scientific evaluation that there are reasonable grounds for concern that the potential dangerous effects on the environment, human, animal or plant health may be inconsistent with the chosen level of protection. It is not intended as an alternative to the risk-based approach, but to guide action when the state of knowledge or the uncertainties in presence do not allow a full scientific assessment of risks, but there are indications of a risk incompatible with the chosen level of protection. The Precautionary Principle is mentioned in various international conventions and agreements, with different formulations. Nevertheless, it has been a subject for controversy between jurisdictions and debate among scholars and policy researchers.
6. The comparative assessment of risk management options often requires the establishment of a risk-risk balance, when a measure addressing a certain risk involves the appearance of an ancillary risk of the same or a different nature. However, ancillary risk reduction benefits are also possible in certain cases. Hypothetical ancillary risks are sometimes invoked to oppose measures addressing well-established risks. In fact, depending on the nature of the risks, the structure of the measure and the regulatory background, it is often possible to identify risk management approaches with a net risk-reduction profile.

7. Monetary risk-benefit assessment poses specific problems when applied to health and environmental risks. Although approaches have been developed for the monetary assessment of human life, damage to health, value of environmental goods and natural resources, both conceptual and ethical problems remain in that respect. Therefore, additional indicators have been developed like for example "quality adjusted life years" to quantify the health-related aspects of risk and the benefits of protection measures.
8. Although the risk-based approach is the "golden standard" of risk policy and regulation, there are cases where measures must be taken on consideration of hazards. That is notably the case of uses of dangerous substances for which there is no practical means to control human exposure or the dispersion and fate across environmental media.
9. Assessing the effectiveness of risk reduction measures requires the establishment of criteria for the acceptability of residual risks. Various approaches have been developed to grade risks in a technical perspective. However, in that respect, a wider appraisal is necessary in order to factor in societal aspects through an open, participatory governance approach. In certain cases, notably for the risks related to exposure to the ionising radiation, it is assumed that risk reduction measures should go as far as practically feasible under the ALARA ("as low as reasonably practicable") approach.

References

ECHOUTCOME. (2013). *Validating outcomes in health technology assessment.* Final report, Project 242203.

EFSA-European Food Safety Authority. (2012). *Risk communication guidelines.*

European Commission. (1997). *Consumer health and food safety*, COM(97) 1997.

European Commission. (2000). *Communication on the precautionary principle*, COM(2000)1.

European Commission. (2006). *Better regulation simply explained.*

European Commission. (2009). *Impact assessment guidelines*, SEC(2009) 92.

European Commission. (2015-1). *Better regulation* for better results, COM(2015) 215.

European Commission. (2015-2). *Better regulation guidelines*, SWD(2015) 111.

European Court of Justice. (1998). *Cases C-157/96 and C-180/96.*

European Court of Justice. (2007). *Case T-229/04.*

European Union. (2002). *Regulation (EC) No 178/2000 of the European parliament and the council laying down the general principles and requirements of food law, establishing the food safety authority and laying down procedures in matters of food safety.*

European Union. (2009). *Regulation (EC) 1107/2009 of the European parliament and the council concerning the placing of plant protection products on the market.*

European Union. (2013). *Council directive 2013/59/EURATOM laying down basic protection standards against dangers arising from exposure to ionising radiation.*

IPCC-International Panel on Climate Change. (2009). *Fifth assessment report.*

ISO (International Standardisation Organisation). (2009). *Risk management-principles and guidelines,* Standard 31000:2009, International Standardisation Organisation.

SCCS, SCHER, SCENIHR. (2013). *Making risk assessment more relevant for risk management.*

Sunstein, C. R. (2003). *The paralyzing principle.* Regulation, Winter 2002–2003.

United Nations. (1992). *Rio declaration on environment and development.*
United Nations Secretariat of the Convention on Biological Diversity. (2000). *Cartagena protocol on biosafety to the convention on biological diversity.*
US 1980. (1980). *The comprehensive response, compensation and liability act (CERCLA).*
US NRC (National Research Council). (1983). *Risk assessment in the federal government: managing the process*. Washington, DC: National Academies Press.
US NRC (National Research Council). (2009). *Science and decisions*. Washington, DC: National Academies Press.
Wiener, J. B., et al. (2011). *The reality of precaution: comparing risk regulation in the United States and Europe*. Washington, DC: RFF Press.

Chapter 6
Risk Communication and Risk Dialogue

Abstract Risk communication serves different objectives depending on the specific circumstances of the risk considered and the aims of communicators. Under emergency and crisis conditions the aim is to provide timely, honest and useful information to the public notably in order to enable potentially exposed people to protect themselves, reduce the uncertainties and the deriving anxiety and stress and prevent behaviour that might aggravate the risks or increase the damage or costs incurred. In such cases, the information is mostly unidirectional, but collecting feedback is important in order to assess how the information is received and understood. Modern risk communication as part of the regulatory process is instead conceived as a two ways process involving in a sustained dialogue, through appropriate stages and procedures, all the relevant partners, managers, assessors, stakeholders, the public and decision makers. The objective is to adequately inform the decision-making process, integrate the societal dimensions of risk into the appraisal, ensure transparency and accountability and prepare the conditions for the effective implementation of risk management measures. The EU has set out procedures for stakeholder and public consultation, during both the risk assessment and the risk management stages. In particular, EFSA and the Commission Scientific Committees have internal rules on public consultation at various stages of their assessment activities. Moreover, the Commission has established consultation principles and guidelines, as part of its Better Regulation guidelines, which apply to all major initiatives including regulatory proposals and decisions on risks.

6.1 Introduction

Risk communication is one of the three pillars of the risk analysis paradigm. Over the years and with the consolidation of risk analysis practices across jurisdictions and sectors, the meaning, the intended aims and the practical operation of risk communication have significantly changed. Best-practice risk communication is no longer conceived as just a one-way flow of information to policy makers, stakeholders and the public from assessors and/or managers, intended to "explain" or "translate" in lay terms the scientific findings of technical risk assessment. In spite of the continued use of the term "communication", a best-practice process is or should be conceived as a two-ways sustained and structured dialogue between all the actors

B. Delogu, *Risk Analysis and Governance in EU Policy Making and Regulation*,
DOI 10.1007/978-3-319-30822-7_6

and across all the stages of the risk analysis process. Still, misunderstandings persist in the way risk communication and risk dialogue are conceived by both policy makers and the public. And often the very objectives of communication are not clear to both sides. The temptation to deploy communication strategies aimed to influence or "convince" the public is still alive…and most often frustrated. While it may be legitimate to try to induce safer behaviours or healthier life styles through targeted communication, while respecting freedom of choice, manipulation strategies and paternalistic attitudes are counterproductive and incompatible with the objectives of best-practice risk governance.

Modern risk communication must reflect the wish and expectation of people in advanced societies to have a say on their own health and safety. In that respect, this Chapter on risk communication and dialogue should be read as an introduction to what we will discuss in the next one on risk governance and a participatory approach to risk management.

6.2 The Different Objectives and Characteristics of Risk Communication

Risk communication is practiced under very different circumstances and with different objectives. Traditionally, several objectives are assigned to risk communication: general information and education of the public on different categories of risks; advice on risk prevention or risk reduction behaviour; warnings and safety advice or instructions in case of emergencies; joint appraisal, problem solving and conflict resolution. It is therefore necessary to introduce some important distinctions. Here we need to make in particular a clear distinction between the risk communication relating to emergencies or crises, which is aimed at providing information and/or safety advice to populations potentially exposed to natural or technological risks, and communication and dialogue as a component of risk analysis, which is part of regulatory and governance processes.

6.2.1 Emergency and Crisis Communication

Risk communication provided to the public in case of emergencies or crises is in the first instance a responsibility of the authorities in charge of the protection of public health, safety and the environment. The aim is to provide timely, honest and useful information to the public notably in order to enable potentially exposed people to protect themselves, reduce the uncertainties and the deriving anxiety and stress and prevent behaviour that might aggravate the risks or increase the damage or costs incurred. EU institutions may be involved in managing health and safety crises and communicating risks related to dangerous products regulated at EU level, for example food, pharmaceutical products, medical devices, or in emergency management related to cross-boundary epidemics or pandemics, where the EU co-ordinates and supports Member States' action. Because of its very nature, risk communication in case of crises or emergencies is mainly a one way process. Nevertheless, in order to check and improve the effectiveness of the information provided it is very important to collect, analyse and respond to feedback from the receivers.

Emergency/crisis communication at EU level poses two main practical problems. First, collecting up-to-date, reliable and verifiable information, and timely monitoring developments during a crisis may be particularly difficult for the EU institutions and agencies because they depend for their own information on the input provided by the large variety of national, regional or local authorities, companies and operators that may be involved in managing a cross-border, regional or global crisis. Second, ensuring a common and coherent position and a consistent, appropriate and effective message for the public across the entire EU is a challenging task. There is no such thing as a "European public". Across the internal EU borders, the public is characterised by significant cultural and language differences, interests, values and expectations. Moreover, account must also be taken of the plurality of actors and the multiple levels of responsibility (European, national, regional, local) across the EU and its Member States. Different responses to a crisis and even slightly different messages seriously undermine the effectiveness of crisis management and the credibility of the information provided to the public.

Risk communication is notably practiced as part of preparedness exercises on emergency management of natural events or technological accidents. This type of communication can be better managed at local level. In fact, it requires a sustained and targeted dialogue with, and participation by the local population, communities or groups at risk.

Risk Communication According to WHO

Risk communications should be a multi-level and multi-faceted process which aims to help stakeholders define risks, identify hazards, assess vulnerabilities and promote community resilience, thereby promoting the capacity to cope with an unfolding public health emergency. An essential part of risk communication is the dissemination of information to the public about health risks and events, such as outbreaks of disease. For any communication about risk caused by a specific event to be effective, it needs to take into account the social, religious, cultural, political and economic aspects associated with the event, as well as the voice of the affected population. Communications of this kind promote the establishment of appropriate prevention and control action through community-based interventions at individual, family and community levels. Disseminating the information through the appropriate channels is also important.

Communication partners and stakeholders in the country need to be identified, and functional coordination and communication mechanisms established. In addition, it is important to establish communication policies and procedures on the timely release of information with transparency in decision making that is essential for building trust between authorities, populations and partners. Emergency communications plans need to be developed, tested and updated as needed (WHO 2011).

6.2.2 *Risk Dialogue with Local Communities*

Risk communication and dialogue takes also place with local communities that may be affected by projects or activities with a potential local impact, like for example the building or operation of power plants, transportation infrastructure, waste treatment facilities or other industrial plants. Such developments and activities are often very controversial. Even for activities on which there is a generic consensus, the NIMBY syndrome (Not In My Back Yard) makes it most often difficult to ensure acceptance of a specific project on the local level. In such cases, the objectives of communication are to identify mutual solutions and respond to concerns, taking into account the right of the local public to be fully involved in decisions of a direct interest for them. Specific guidelines have been developed for community communication. For instance, the EPA has identified *Seven Cardinal Rules of Risk Communication* in that respect (EPA 1988):

1. Accept and involve the public as a legitimate partner. A basic tenet of risk communication in a democracy is that people and communities have a right to participate in decisions that affect their lives, their property and things they value.
2. Plan carefully and evaluate your efforts. Risk communication will be successful only if carefully planned.
3. Listen to the public's specific concerns. If you don't listen to people you cannot expect them to listen to you. Communication is a two-way activity.
4. Be honest, frank, and open. In communicating risk information, trust and credibility are your most precious assets.
5. Coordinate and collaborate with other credible sources. Allies can be effective in helping you communicate risk information.
6. Meet the needs of the media. The media are a prime transmitter of information on risk; they play a critical role in setting agendas and in determining outcomes.
7. Speak clearly and with compassion. Technical language and jargon are useful for professional shorthand. But they are barriers to successful communication with the public

Risk communication for crisis or emergency preparedness and management or for local projects and activities is defined and organised in view of very peculiar aims and is somewhat outside the conceptual paradigm of regulatory risk analysis. We do not address the details of crisis or emergency management in this book, The focus of the following sections is instead specifically on risk communication as part of the EU risk regulatory processes.

6.3 Risk Communication in the EU Regulatory Process

6.3.1 The Objectives and Contents of Risk Communication in the Regulatory Process

Risk communication as part of the regulatory process is aimed at establishing and conducting a sustained dialogue across the decision making chain and with all the relevant stakeholders and the interested public. Such dialogue serves various complementary objectives.

The first objective is supporting the identification, understanding, definition and framing of the relevant risk issues. To that aim, it is important in the first instance to promote and support a sustained technical and scientific dialogue, complemented by data exchange and analysis, among the relevant scientific and technical bodies, in the EU and internationally. Such scientific dialogue helps identifying new and emerging risks, notably by scanning the technological horizon and by monitoring and assessing the scientific developments that may have an interest for the early identification and assessment, or re-assessment, of risks. In the EU, the dialogue among national experts, the EU and national technical and scientific bodies and the Member States' authorities is of particular importance in order to ensure the sharing of knowledge and for promoting consistent views on the relevant risk issues among the scientific and institutional actors.

The experience shows that risk issues may become politically important even if they have not initially been considered a priority from a purely scientific and technical point of view. Risk perception often plays a dominant part in setting the political agenda. It is therefore important for the early identification of sensitive risk issues to establish and maintain a stakeholder dialogue with the aim to help exploring and understanding public perception of new, emerging or anticipated risks. The objective is to frame the early assessment of actual or potential risk-issues within the broader perspective of stakeholders and the public. Moreover, risk-issues are often addressed at EU level when they have already become controversial and emotional, sometimes after events or cases of true or alleged mismanagement have caused public outrage. In such cases, it is the outrage more than the hazard that needs to be addressed. Addressing the opinions, emotions and reactions of the public, managing joint problem-solving and conflict resolution processes requires specific skills. While scientists and managers need to be involved in the dialogue, it is important that professional communicators with good listening skills help structuring the dialogue process and support it.

Discussion and consultation across the EU institutional levels and with the stakeholders are key for formulating the questions to be submitted to risk assessors in a way properly addressing both the technical aspects of risks and the concerns of stakeholders and the public, with a view to ensure that the scientific advice delivered by the assessors effectively supports decision making deliberations. Risk managers should also seek feedback on draft risk assessment mandates from the risk assessors themselves before finalising their questions for the scientists.

It is important to maintain open communication between risk managers and risk assessors along the entire assessment process, with the objectives of optimising the utility of the scientific advice in relation to the needs of decision makers, of ensuring that the risk assessment policy-issues are timely and transparently addressed by managers and assessors in collaboration, and for ensuring clarity, transparency and proper understanding of all the relevant assumptions, defaults, uncertainties intervening in the assessment.

The responsibility for selecting and weighing the data and information to be used in risk assessment remains with the scientists. Nevertheless, it is important to consult broadly with the scientific community, the stakeholders and the public at an early stage of the risk assessment process in order to collect all data and information that may be relevant or may be proposed for consideration. Similarly, while risk assessors must remain independent in deciding their conclusions, it is good practice to offer stakeholders the opportunity to comment on a preliminary version of the risk assessment opinion and to explain how comments have been considered and, if appropriate, taken into account.

The overall aim of risk communication and risk dialogue at the assessment stage is to ensure the best conditions for providing the most policy-relevant, high quality, transparent, understandable and accepted scientific opinion.

A well designed communication and dialogue approach should also address the assessment of the risk management options as well as the final regulatory measures. It is again important to engage with all the actors and interested parties at that stage, as part of the regulatory impact assessment process. Although the result of the risk assessment plays a most important part in deciding the appropriate course of action, the stakeholder feedback from a broad consultation helps the necessary multi-dimensional and multi-criteria appraisal of the policy options leading to an effective, more widely accepted, or at least better understood, political decision. Finally, it is good practice to maintain open the risk dialogue after measures are decided, at their implementation stage, in order to monitor how the decisions are received and applied, what are their effects, what is achieved, and to identify timely any possible need for revision or improvement.

6.3.2 Organisational and Practical Aspects of Risk Communication

Effective risk communication requires a substantial organisational effort. The starting point is to establish clearly defined operational procedures for consulting and discussing with actors, stakeholders and the public. This is a complex task on the EU-level with several institutional layers involved, 28 Member States, each with its internal administrative, technical and scientific structures, a variety of stakeholders with diverse interests, and public opinions that are far-away from the EU Institutions and are characterised by different languages, cultures, values, interests and risk perceptions. For a start, there is a need for offering easily accessible information, provided in recognisable web-sites, covering the working procedures, programs, initiatives, consultations and opportunities for interaction and stakeholder dialogue. The "rules of the game" need to be stated clearly, specifying what are the objectives and limits of the dialogue, how comments are taken into account and how decision-making responsibilities are organised. The experience shows that it is counterproductive to create excessive expectations. It should be clearly stated that a dialogue does not imply a sharing of responsibility for decisions, and that the dialogue procedures should not be misused for imposing endless debates with the aim of unduly delaying decisions.

Modern information and communication technology offers powerful means for reaching a variety of audiences and for collecting feedback. Nevertheless, the effective appraisal of concerns, positions and perceptions may require in-person dialogue and the use of techniques like focus groups, hearings, workshops etc.

A particular effort needs to be done in order to ensure, as far as feasible, representativeness and balance among stakeholders in the consultation and dialogue processes. There is an evident asymmetry in the representation capacity between the organised industrial, professional, economic interests and other parts of the civil society that are unevenly represented by NGOs. NGOs are often characterised by narrow sectorial interests, limited resources and lack of technical expertise. Their public statements may be strident, but sometimes their weight in the policy debates suffers from their lack of technical expertise. Because of such weaknesses, NGOs may be tempted to leverage their action on the emotional feelings of the public, knowing that they might lose the case in a purely factual and technical debate, account taken of the disproportion between their resources and those mobilised by the much better organised economic and industrial interests. It is therefore important to support the full involvement of the weaker stakeholders in a factual debate, enabling them to provide their input on all aspects, including the scientific ones.

Finally, a considerable effort is required for ensuring an adequate follow up to the consultations, notably the analysis of the contributions received and of the positions expressed. A fair balance in considering the opinions expressed by the various parties, and transparency on the outcome of consultations should always be ensured. Feedback should be given on how the contributions, comments and criticisms received have been considered, taken or not taken into account, and why and how they are possibly reflected in the final decisions.

The success of risk communication and dialogue depends heavily on the trust in and credibility of the institution or body conducting the process. Once again, honesty and clarity on the objectives and limits of the dialogue are vital. Misunderstandings on the objective of a risk dialogue may cause frustration, compromising the trust in the institution and the interest of stakeholders in the process. Trust of the participants depends in the first instance on the clear commitment to the process by the leading organisation. Conducting a consultation as nothing more than a formal procedural requirement, not expected to add much to the scientific risk assessment and to the impact assessment studies, is a sure way to lose all credibility with the stakeholders and the public. Commitment, competence and appropriate communication skills, adequate resources, respect for and balanced consideration of all the opinions expressed, are all required features for the credibility of communication, consultation and dialogue initiatives.

6.3.3 The Interlocutors

We have seen that risk communication has been interpreted in the past (and is sometimes still misunderstood) as a process where the authorities and the experts are the providers of information, while the stakeholders and the public are on the receiving end. We have noted that risk communication is nowadays intended as a structured dialogue that must involve risk assessors, risk managers, decision makers, stakeholders and the general public. The final objective of such a dialogue is to determine the best policy or regulatory decision, addressing all the dimensions of the risk issue considered, and to contribute to ensure the effective implementation of the measures taken.

On the EU institutional level, the relevant interlocutors and actors of the risk dialogue are well defined: the EU risk assessment bodies competent for providing scientific advice on the relevant subject matter, the Commission with its internal administrative articulations, the European Parliament, and the Council with the experts from the competent authorities and technical bodies of the Member States. So far, the rules and practices of the EU decision-making process have not integrated a sustained inter-institutional dialogue across the various stages of risk assessment and management. For instance, it has not been common practice of the Commission to involve the Parliament and the Council in defining the scope of a risk assessment and in deciding on the questions for the scientists. Nevertheless, the recently reinforced "Better Regulation" rules could provide a framework for a closer collaboration between the Institutions since the early stages of risk analysis.

On the side of EU stakeholders, a wide range of civil society interests are represented to the EU through industrial and professional associations or confederations, trade unions, NGOs etc. Consumers, patients and various categories of citizens also have their own European organisations. And a variety of thematic NGOs follow developments of EU policies on the environment, climate change, energy etc.

Finally, many non-EU countries that may be affected by the EU risk decisions are also represented to the EU Institutions by their diplomatic missions.

6.3.4 Current EU Practice

Under the current EU practice, risk communication is mainly managed by the European Commission and the EU risk assessment bodies, in particular EFSA, EMA, ECHA and the ECDC. Risk communication, consultation and dialogue on risk management and regulation takes place within the framework of the Better

Regulation procedures, which include provisions on consultation standards. The Commission has established a variety of procedural steps in order to involve the main interested actors and stakeholders in the preparation of proposals and decisions on policies and measures, over and above the institutional procedures of the Treaties. Some of those procedures apply in a systematic way while others are applied on an ad hoc basis. While such procedures are not specific to risk measures, they also apply to decisions concerning risks and provide an effective framework for risk dialogue. Green Papers and White Papers are published in order to communicate and consult on preliminary analyses of policy issues and policy options. On an ad hoc basis, conferences and workshops are organised in order to monitor and discuss particularly important and sensitive risk issues. Finally, when policy or regulatory measures are considered, the Better Regulation rules provide for a systematic, public consultation on policy options in accordance with set standards. Impact assessment reports are made publicly available in the Commission web-site. The minimum standards for consultation adopted by the Commission represent a very comprehensive framework for effective risk dialogue (European Commission 2015):

General Principles and Minimum Standards for Consultation

Relations with stakeholders are governed by four general principles:

1. Participation: Adopt an inclusive approach by consulting as widely as possible;
2. Openness and Accountability: Make the consultation process and how it has affected policy making transparent to those involved and to the general public;
3. Effectiveness: Consult at a time where stakeholder views can still make a difference, respect proportionality and specific restraints;
4. Coherence: Ensure consistency of consultation processes across all services as well as evaluation, review and quality control.

These principles are complemented by five Minimum Standards that all consultations have to respect:

(A) Clear content of the consultation process ('Clarity'): All communication and the consultation document itself should be clear, concise and include all necessary information to facilitate responses;

(B) Consultation of target groups ('Targeting'): When defining the target group(s) in a consultation process, the Commission should ensure that all relevant parties have an opportunity to express their opinions;

(C) Publication: The Commission should ensure adequate awareness-raising publicity and adapt its communication channels to meet the needs of all target audiences. Without excluding other communication tools, (open public) consultations should be published on the internet and announced at the "single access point";

(continued)

(D) Time limits for participation ('Consultation period'): The Commission should provide sufficient time for planning and responses to invitations and written contributions;
(E) Acknowledgement of feedback ('Feedback'): Receipt of contributions should be acknowledged and contributions published. Publication of contributions on the "single access point" replaces a separate acknowledgment if published within 15 working days. Results of (open public) consultations should be published and displayed on websites

The EU risk assessment processes also include procedures for the consultation of experts, stakeholders and the public. Such procedures are part of the rules, procedures and practices of the EU risk assessment bodies and Scientific Committees. Two important examples are provided by the rules of procedure of the EC Scientific Committees and by the procedures which are applied by EFSA. The main elements of the stakeholder dialogue procedure of the Commission Scientific Committees (EC Scientific Committees 2009) are reported hereafter as an example of good practices allowing for openness and dialogue while safeguarding the integrity and independence of the scientific process, anticipating on the content of Chap. 8. Stakeholders interested in contributing to the dialogue on particular themes find all the necessary information in the websites of the relevant risk assessment bodies.

The Stakeholder Dialogue Procedure of the Commission Scientific Committees

The procedure is aimed "to enable structured, balanced, ordered and manageable engagement with stakeholders in the process of elaboration of scientific opinions by the Commission Scientific Committees". The procedure is to be applied in a way preserving the independence of the Committees. The overall aim is "to contribute to ensure the highest quality of the scientific opinions adopted by the Committees". The commitment to consult is subject to consideration of the resources available and the expected added value of the dialogue in each specific case.

The following aspects of the Committees' activities are covered by the procedure:

– suggestions for new topics which the Commission may consider to submit to a Scientific Committee: proposals for new issues that fall under the competence of the Committees may be considered provided several conditions are met, notably if they concern clearly formulated scientific risk assessment questions, have a specific interest for the EU and are supported by relevant data and scientific knowledge;

(continued)

- finalisation of new mandates: all new mandates for the Scientific Committees are published on the Internet. 'Working' mandates are submitted to public consultation in the case of "issues of broader significance or wider public interest";
- calls for data and information: such calls are published when the Committees consider it useful to collect widely relevant data and information, with the aim "to ensure that all relevant scientific information as specified in the Call is available to the Scientific Committee for its assessment";
- scientific input during the preparation of the opinion: this may include technical hearings with individuals, petitioners or other stakeholder representatives, open to scientists with appropriate expertise, and on occasion, open public hearings to gather specific comments, suggestions, explanations or contributions on the scientific basis of a particular opinion.
- public consultations on a "pre-consultation" opinion: preliminary versions of certain opinions are submitted to a public consultation in order to gather specific comments and suggestions on the relevant scientific aspects with a view to contribute to the quality of the opinions.
- revision of existing opinions: the revision of an existing opinion is initiated when new evidence become available showing the need to modify or update it, notably when new data and information is provided at the request of the Scientific Committees or at the initiative of stakeholders, international organisations, third countries, or when evidence showing previously unidentified hazard properties, exposure situations, or potential risks is submitted by Member States notifying to the Commission measures restricting the marketing of products regulated by EU internal market legislation.

Similarly, EFSA has a structured procedure for consultation covering the scientific opinions and the statements of EFSA Scientific Committee and/or Panels, the statements of EFSA, the guidance documents of the Scientific Committee, Panels and EFSA, the scientific or technical reports of EFSA and EFSA self-task outputs.

As in the case of the Commission Scientific Committees, EFSA's policy on consultation is aimed both at ensuring transparency and accountability, and at contributing to a high level of scientific quality and efficiency. Openness is particularly crucial in the area of food safety since one of the major aims of the creation of EFSA was to regain consumer confidence in the EU food safety system after the BSE crisis (EFSA 2012).

Finally, other EU agencies with a technical role, like EMA, are less involved in a stakeholder dialogue on a routine basis. However, all of them have web-sites where information on their activities can be found, including on events and opportunities for presenting comments on an ad hoc basis.

6.3.5 Limits and Problems of Risk Communication

The quality and success of risk policies and regulatory decisions depend to a great extent on effective risk communication and dialogue. However, risk dialogue is a challenging endeavour, requiring adequate resources and an appropriate, shared institutional culture. Openness and dialogue are aspects of EU decision-making practice that are well developed and continuously improving. However, risk dialogue has its own drawbacks, limits and problems. We highlight here some critical aspects.

Delaying Tactics The main risk of excessively open-ended consultation practices is that of offering opportunities to vested interests for unduly influencing or delaying decisions. Experience shows that not all the interested parties are prepared to engage constructively and loyally in a productive dialogue. Clarity and firmness on the rules are essential in order to discourage disruptive or improper behaviour.

Misunderstandings on the Nature of Risk Dialogue Sometimes the nature of the risk dialogue is misunderstood, notably when excessive expectations are raised. Certain stakeholders may be misled to believe that consultation initiatives open the way to participation in decision making. They are deceived, and become critical and sceptical about the usefulness of dialogue processes, when in the end the decisions taken by the Institutions do not reflect their contributions or are not in line with their positions. It is important to make clear at the beginning of the process, that consultation and dialogue on risk assessment and risk management options do not involve any negotiation on science or sharing of responsibility for the final policy or regulatory decisions. Consultation and dialogue are part of a participatory approach aimed at *informing* decisions. However, under the EU legal order, decisions are taken in accordance with the decision-making processes of representative democracy.

Unbalanced Representation of Civil Society Sectors In the absence of a proactive support-policy aimed to ensure a fair balance between all the relevant stakeholders, the disproportion of resources favours the most powerful and better organised groups that can participate in the consultation process with data, expertise and well-prepared legal and scientific arguments.

Limited Representativeness of Certain Organisations and Prevalence of the Minimum Common Denominator Further difficulties are related to the limited representativeness of certain sectorial organisations and their need to mediate among the diverse conditions and positions of their members. Within the same sector or group of the civil society, interests may vary substantially across such a differentiated ensemble as the EU. Not all the companies in a given sector are at the same level in terms of innovation, competitiveness, technical, scientific and organisational capacity.

When commenting on policy options, trade associations are sometimes pushed to take positions that reflect the situation of the less innovative and weaker companies. Consumer associations may sometimes be in the difficult position of favouring EU measures that improve the protection in certain countries but are at a lower level than that already prevailing in the few most-advanced ones. And NGOs sometimes communicate a perception of risk that in fact is not shared by the public opinions in all the Member States.

Need for Adequate Resources Finally, the proper management of risk communication is resource-intensive. The scope, rules and procedures set out for risk communication should therefore be commensurate with the resources available.

Need to Balance Completeness/Accuracy and Accessibility of the Information Provided Providing complete and accurate information is part of risk communication. While transparency and access to all the available information and data must be ensured, it is often difficult to balance completeness and accuracy on the one hand with clarity and meaningfulness for the layman on the other hand. Oversimplification may cause misunderstandings and raise suspicion of manipulation. In turn, using rigorous scientific language and presenting a large amount of data may exclude the non-expert from the dialogue.

Excessive Expectations on Consensus Building Through Communication and Dialogue Policy makers and risk managers often expect that their efforts to communicate would lead to a larger consensus across the various stakeholders and the interested public on the relevance and level of risk and on the management measures, preventing or at least reducing conflicts and smoothing decision making. However, that "happy end" does not easily happen. There are many barriers to constructive and productive dialogue. Different groups and stakeholders may perceive in different ways the same risk, its meaning, size, level of priority. They can have different views on the values underlying the appraisal of a risk and on the relevance or reliability of scientific facts. Also, the level of understanding of the scientific aspects may vary significantly. The various groups of exposed people may have very different levels of receptivity to risk messages, depending on their risk-aversion and whether they feel to be more or less directly concerned. Finally, a productive dialogue requires a genuine wish to engage of all the parties interested and some openness to look for and endorse mutually acceptable, compromise solutions. In many cases, a more modest but realistic and worth aim for risk communication and dialogue is to contribute to a better understanding by risk managers of all the relevant aspects of the risk-issue considered, including the non-scientific ones, facilitating the identification of robust risk policies and measures.

Summary

1. Risk communication is the third pillar of the risk analysis paradigm. It used to be conceived as a one-way flow of information from experts and managers to the public, but modern risk communication is nowadays considered to involve (apart

from emergency situations) a two-ways dialogue among all the relevant actors, scientists, managers, policy makers, stakeholders and the public. Such dialogue is intended to ensure transparency and accountability and to contribute to the quality and effectiveness of risk assessment, and of the management measures to be adopted.

2. The content and structure of risk communication depend on the specific circumstances and on the relevant needs and objectives of the case. In the case of emergency or crisis communication, the aim is to provide information enabling potentially exposed people to protect themselves, reduce the uncertainties and the deriving anxiety and stress and prevent behaviour that might aggravate the risk. Feedback from the receivers is important in order to check and improve the effectiveness of the information provided and acquire knowledge on exposure situations. The EU is involved in managing crisis situations and needs to cooperate with its Member States in communicating risks in such cases. There are specific problems in that respect due to the variety of cultures, languages, attitudes to risk and expectations across the EU.
3. Risk dialogue takes place with specific communities in case of projects, development and activities that may have an impact on the local level. There are many examples of developments that become highly controversial, like for example transportation infrastructure, power plants, waste treatment or disposal facilities etc. Developers and the authorities are often confronted with the so-called "NIMBY (Not in My Back Yard) syndrome. Specific guidelines have been developed for community communication. Recognising the legitimacy of local community involvement in the decision making is a precondition for a fruitful dialogue.
4. Risk communication as part of the regulatory process is aimed at establishing and conducting a sustained dialogue across the decision making chain and with all the relevant stakeholders and the interested public. There are several complementary objectives for such a dialogue in its different articulations: supporting the identification, understanding, definition and framing of the relevant risk issues; collecting useful information and data for risk assessment; identifying new and emerging risks; identifying at an early stage risk issues that may become controversial and crystallised; ensuring clarity and understanding on the scientific aspects of risks etc.
5. Effective risk communication and dialogue require significant resources and the application of a variety of procedures and technical means. On the EU level, communication and dialogue are complicated by the multiple layers of the decision making process, the parallel voices of 28 Member States, the different cultures, languages and risk attitudes across the Union. Structured procedures for communication and/or information have been put in place by the Commission, its Scientific Committees and the EU Agencies dealing with risk assessment, notably EFSA. With the reinforced Better Regulation initiatives introduced by the European Commission in 2015, consultation on the EU level covers the entire life cycle of EU regulations.

6. Risk dialogue contributes to the effectiveness and efficiency of risk policy and regulation. Nevertheless, it is a challenging endeavour and has its own potential risks and drawbacks as shown by experience. It may encourage delaying tactics if the rules of the game are not clearly defined and applied. The aims and limits of the dialogue may be misunderstood creating excessive expectations on the involvement in the decision making, which in our systems of indirect democratic representation remains the preserve of political institutions, and resulting in frustration and bitter controversies. The process may in some cases become captive by dominant lobbies able to deploy resources that are not available to NGOs and other sectors of the civil society. And finally, open ended commitment to dialogue in the absence of clear cut-off rules may result in disproportionate administrative burden not justified by the actual added value of the process.

References

EFSA. (2012). *When food is cooking up a storm – proven recipes for risk communications.*

European Commission. (2015). *Better regulation guidelines*, SWD (20015) 111.

SCCS, SCHER, SCENIHR. (2009). *Rules of procedure of the scientific Committees on consumer safety, health and environmental risks, emerging and newly identified health risks.*

US EPA (Environmental Protection Agency). (1988). *Seven cardinal rules of risk communication.* Washington, DC: Office of Policy Analysis.

WHO (World Health Organisation). (2011). *IHR core capacity monitoring framework: Checklist and indicators for monitoring progress in the development of IHR core capacities in states parties international health regulations coordination.*

Chapter 7
Risk Regulatory Policy and Risk Governance

Abstract Although not all risks fall under the responsibility of public authorities, there is a need for risk regulation in cases where individuals have no effective means and powers to protect themselves, or no interest in preventing or reducing risks imposed on others. On the EU level, regulation of health, safety and environmental risks is mostly motivated by the need to harmonise rules in order to ensure the functioning of the internal market. Critics of EU "overregulation" often ignore the enormous simplification brought by such regulation compared to the alternative of a 28-Member States fragmented market. Formal requirements on quality and transparency of EU regulation are not new and both the Treaties and the European Court's case law clarify that all measures must be motivated and that the Union must take into account the available scientific and technical data. Moreover, a good part of the harmonisation legislation is based on a regulatory approach that makes only essential safety requirements mandatory, leaving to industry and other interested parties to set up detailed (voluntary) standards. The European Commission has set a Better Regulation approach in order to ensure the quality of legislation, review and revise existing legislation and ensure transparency and the involvement of stakeholders and the public, across the entire life cycle of regulations. It of the utmost importance to put evidence and science at the core of the preparatory and evaluation process. However, regulation may be "science and evidence-based" only in the sense that evidence and science are of great importance to <u>inform</u> the decision-making process. Other factors play an essential part in the process, and the final outcome is determined by a dialectical, political appraisal and the democratic exercise of institutional power. In order to ensure the effectiveness and acceptance of regulatory measures, the government process must be framed within a broader governance approach, where technical assessment is complemented by a broader concern appraisal and decisions are designed in light of inputs coming from a participatory approach.

7.1 Introduction

This Chapter is devoted to the broader context in which risk analysis, the assessment, management and communication of risks, is framed: the risk regulatory policy and governance, in their technical and institutional dimensions and within the

B. Delogu, *Risk Analysis and Governance in EU Policy Making and Regulation*,
DOI 10.1007/978-3-319-30822-7_7

wider environment of stakeholder and public interests and involvement. We address here both the "governing" and the "governance" of risks.

In this Chapter we talk extensively about "regulation" and regulatory policy. We do not make here a distinction between "legislation" and "regulation". We have already explained in Chap. 5 that the EU has its own particular way to legislate and regulate, with specific procedures and terminology.

The EU legal instruments are binding regulations, directives and decisions and non-binding recommendations, which are adopted by the European Parliament and the Council (the EU legislator) and, under certain conditions, by the Commission, notably based on delegated or implementation powers attributed to it by the EU legislator. While in some other jurisdictions the term "regulation" usually indicates a secondary level of rules, made by an executive authority under powers delegated from the legislature, detailing the requirements aimed at the implementation of primary, more general legislation, this characterisation does not apply to EU "regulations" which are established by the EU legislator and often set up wide legislative frameworks, like for example in the case of the REACH Regulation on chemicals. EU regulations and directives may include both general requirements and more detailed provisions or standards. Anyway, for the aims of this Chapter we do not need to make a distinction between regulation and legislation or between the different types and level of legal instruments. We are in fact interested here in discussing the characteristics of the processes through which are taken decisions on policies and measures of any type aimed at addressing risks.

In Europe, there is a lively debate on regulation, notably on risk regulation. Often, the EU is criticised for being a source of excessive regulation, compromising the competitiveness of industry and economic activities in general, putting a brake on innovation, creating unjustified barriers to international trade. The rationale, effectiveness, efficiency, scientific justification and sometimes the very need for EU regulation are often challenged. In parallel to the ideological debate on de-regulation, less state or less Europe, more practical considerations and criticisms are raised on the use of science and evidence in the EU regulatory process, the choice of regulatory instruments, the balance between costs and benefits, the flexibility of regulatory requirements and standards, their adaptability to new scientific knowledge and their compatibility with innovation and international competition.

The EU is a Union largely built on the rule of law and on common rules in a variety of domains, notably rules aimed to establish and maintain the four basic "freedoms" of circulation: for goods, services, capital and people. It is therefore not surprising that the quality, effectiveness, efficiency and justification for the substantial body of rules on which the EU is based are constantly scrutinised and debated. And the issue of reducing the regulatory burden is very high on the agenda of the EU Institutions, in the first instance the European Commission, responsible for initiating the regulatory process. In that respect, we have already mentioned the principles and procedures established and gradually reinforced and completed by the Commission within the framework of the Better Regulation approach and we say more on that in this Chapter.

7.2 Issues, Problems, Principles and Practice of Risk Regulation

7.2.1 The Rationale for Risk Regulation

The need and justification of rules established and enforced by public institutions in order to protect rights and legitimate interests, like in particular the right to health and safety, and ensure the ordered functioning of economic activities, prevent and repress disruptive behaviour and, more in general, ensure the many vital general public interests in our complex societies, are so obvious that there is no need for discussing them here. Nevertheless, in the case of risk regulation, we need to make a few specific observations.

In previous Chapters we have highlighted that risks are part of life and that managing the variety of risks with which we are daily confronted is in the first instance an individual responsibility. Nevertheless, there are many situations and categories of risks that cannot be expected to be managed on the individual level, because individuals have no effective means and powers to protect themselves and the goods they value, or no direct, individual interest to prevent such risks, or they even may sometimes get a benefit from creating a risk which is imposed on others. In particular, protecting the environment and the so-called "commons" requires a collective action that can only be set up and enforced by public institutions. The objective of ensuring the proper level of product and service quality and functionality, responding to consumer demand, is in general met through the market and competition mechanism. Consumers may well choose to buy less performant products or services based on cost considerations. On the contrary, in the case of safety, in general it is not left to the supply and demand market-mechanism to ensure the appropriate required minimum level of safety (that in the EU must be "high", according to the provisions of the Treaties).

In some very specific cases, mandatory requirements are also established in order to prevent individual behaviours or activities involving risks for those who could decide on their own to expose themselves to such risks, like for example the establishment of no-bathing or no-skiing zones. Such prohibitions, or more often warnings, are in particular set for locations or situations where a high frequency of dangerous behaviours and accidents may be expected.

7.2.2 Risk Regulation in the EU: Rationale and Objectives

In Chap. 3 we have highlighted the extensive competences of the EU for regulating health, safety and environmental risks. Often the EU is criticised for producing a huge amount of rules, in too many policy areas, with an excessive degree of detail and intrusiveness and little consideration of impacts on innovation, competitiveness and trade. However, many commentators do not seem to take into account the underlying rationale for the regulatory role of the Union. The EU has been defined as a "regulatory polity" (Majone 1996). In fact, after some failed attempts to create a political union in one go at the beginning of the 1950s, the strategy of the founding fathers and the European Institutions was to build and progressively expand and reinforce the Union through the development of functional, sectoral communities, promoting the solidarity, collaboration and integration between Member States in key areas. The programme of functional integration was already clear in the Schuman declaration of 9 May 1950 announcing the creation of the first European "Community", the European Community of Steel and Coal:

> Europe will not be made all at once, or according to a single plan. It will be built through concrete achievements which first create a de facto solidarity.

The most important vehicle of integration was in fact established in 1957 with the creation of the European Economic Community, notably its internal market and the common agricultural policy.

The internal market was and still remains the backbone of the Union. It has been progressively established and reinforced, and is still being gradually completed and extended to services (including financial services, notably with the recent Banking Europe initiatives), through the harmonization of the conditions applicable to the placing on the EU market of goods, to the provision of services, and to the free circulation of capitals and workers. The harmonization of the rules for the establishment and functioning of the internal market has involved a massive regulatory activity. Although the principle of mutual recognition of national standards is part of the approach to removing barriers to internal trade, its practical application has been limited by the recognition, confirmed by the EU Court of Justice, of the right of each national jurisdiction to fix its preferred level of protection and make its own regulatory choices in light of the specific national needs and circumstances. Much of the EU harmonization rules address health, safety and environmental risks, the interests of consumers and fair, undistorted competition conditions for producers and service providers. Apart from legislation on environmental protection, which in general is framed within the specific EU policy on the environment, the justification and added value of most of the EU risk regulation must therefore be assessed in the perspective of the functioning of the internal market. This observation has important consequences.

First, the somewhat inflated discourse on the EU regulatory "red tape" should take into account the need for and added value of a single set of rules for 28 Member States. Less EU regulation would not necessarily lead to less regulation in Europe, with 28 Member States that, according to the doctrine of the European Court of Justice, in the absence of EU harmonization would be entitled to regulate in so far as necessary in order to satisfy mandatory requirements such as for example the protection of health and safety. In such cases, obstacles to the movement of goods or restrictions on the provision of services resulting from disparities between the national regulations ought to be accepted and would involve huge negative consequences for the EU Member States' economies. As far as the quality of the regulation is concerned, it is doubtful that science and evidence, as well as the variety of inputs from the diverse EU interests, would be better incorporated into the regulatory process on the national level, with less capacity in many of the EU countries to mobilize scientific and technical resources and little opportunity and motivation for a comprehensive consideration of the wide range of interests and values at stake across the EU. In terms of innovation and competitiveness, it seems quite obvious that a single set of rules for access to a market of more than 500 million consumers creates much more favourable conditions compared to a proliferation of national requirements. In that respect, the EU harmonisation represents a formidable advantage for producers and service providers, to say nothing of the indirect benefits from the international influence of EU standards.

Second, the regulatory activity for the needs of EU harmonization has in practice transferred to the EU Institutions the responsibility for many aspects of health, safety and consumer protection. In its action to establish and ensure the functioning of the internal market, the EU has been therefore brought to set out continent-wide aims and principles of action on health and safety protection. Risk regulation at EU level facilitates the achievement of a consistent level of protection across the 28 Member States (under however the important, but unfortunately difficult to achieve, condition of effective and consistent implementation and enforcement at national level!).

The EU risk regulatory action finds however its limits in the subsidiarity principle (which is relevant in the area of technical rules since the internal market remains a shared competence). The Member States are notably allowed to legislate in the absence of EU rules, and even to "gold-plate", under certain conditions, EU harmonisation provisions. Moreover, the Member States are competent for the enforcement of EU rules. Finally, in the absence of a legal base in the Treaties for regulating on the EU-level risks to public health unless it can be shown that national rules create obstacles to the functioning of the internal market, the Member States retain the exclusive regulatory competence on public health matters. For example, only a non-binding EU recommendation addresses the sensitive issue of protection of the public from exposure to non-ionising electromagnetic radiation.

The European Commission has clearly stated its views on the rationale for the EU regulation in its communication of 2015 on Better Regulation (European Commission 2015-1). While introducing reinforced and improved criteria and procedures for ensuring the quality of regulation, it has dispelled any concern on alleged "de-regulatory" orientations by stating about the accumulated and new EU legislation that:

This legislation is essential for sustainable development, for the single market that drives our economy and for unlocking the investments needed to support jobs and growth. It underpins the European social model and gives meaning to the rights and freedoms that citizens cherish, including their security and right to justice. It also helps us respond to common challenges such as energy security and protecting the environment and the climate. In many cases, one set of EU rules replaces a patchwork of 28 different national rules, so making life easier for citizens and businesses, simplifying the legal framework, reducing regulatory burdens across the single market and increasing regulatory predictability. The body of EU law is not only necessary, it is our great strength – it makes the EU qualitatively different from any other model of collective governance in the world. That is why it is so important that every single measure in the EU's rulebook is fit for purpose, modern, effective, proportionate, operational and as simple as possible.

In summary, the rationale and objectives of the EU risk regulation are well represented in the Food Law (European Union 2002) and in the REACH Regulation (European Union 2006):

The EU Food Law: Rationale and Aims

Preamble

1. The free movement of safe and wholesome food is an essential aspect of the internal market and contributes significantly to the health and well-being of citizens, and to their social and economic interests.
2. A high level of protection of human life and health should be assured in the pursuit of Community policies.
3. The free movement of food and feed within the Community can be achieved only if food and feed safety requirements do not differ significantly from Member State to Member State.
4. There are important differences in relation to concepts, principles and procedures between the food laws of the Member States. When Member States adopt measures governing food, these differences may impede the free movement of food, create unequal conditions of competition, and may thereby directly affect the functioning of the internal market.
5. Accordingly, it is necessary to approximate these concepts, principles and procedures so as to form a common basis for measures governing food and feed taken in the Member States and at Community level.

…

Article 1

Aim and scope

1. This Regulation provides the basis for the assurance of a high level of protection of human health and consumers' interest in relation to food, taking into account in particular the diversity in the supply of food including traditional products, whilst ensuring the effective functioning of the internal market. It establishes common principles and responsibilities, the means to provide a strong science base, efficient organizational arrangements and procedures to underpin decision-making in matters of food and feed safety

…

The REACH Regulation: Rationale and Aims

Preamble

1. This Regulation should ensure a high level of protection of human health and the environment as well as the free movement of substances, on their own, in preparations and in articles, while enhancing competitiveness and innovation. This Regulation should also promote the development of alternative methods for the assessment of hazards of substances.
2. The efficient functioning of the internal market for substances can be achieved only if requirements for substances do not differ significantly from Member State to Member State.
3. A high level of human health and environmental protection should be ensured in the approximation of legislation on substances, with the goal of achieving sustainable development. That legislation should be applied in a non-discriminatory manner whether substances are traded on the internal market or internationally in accordance with the Community's international commitments.

…

Article 1

Aim and scope

1. The purpose of this Regulation is to ensure a high level of protection of human health and the environment, including the promotion of alternative methods for assessment of hazards of substances, as well as the free circulation of substances on the internal market while enhancing competitiveness and innovation.

…

7.2.3 The EU Risk Regulatory Policy: General Aspects and Principles

We have already illustrated in Chap. 5 the general features of the EU regulatory processes and discussed the general principles established by the EU Treaties, subsidiarity, proportionality and the precautionary principle, and their relevance in relation to risk regulation. We have seen that the multi-level and multi-actor EU decision-making processes, involving institutions that bring together representatives of 28 countries, pose specific principle and practical problems when it comes to ensure a consistent implementation of decision and governance practices. We want here to complete our discussion on the EU "risk government" processes, in particular highlighting the role played in it by science and evidence. In addition to the changes to the decision-making procedures which have been introduced with the successive modifications of the Treaties, the way in which EU regulation is prepared and assessed has been subject over time to important procedural and substantive improvements. Although the focus on systematic impact assessment, transparency and open and inclusive consultation has been enhanced at the beginning of the first decade of this century, it would be wrong to think that older policy and regulatory decisions were established in a vacuum of principles and practices, including consideration of the relevant evidence and scientific data, or without any form of ex-ante evaluation and consultation. All EU mandatory legal acts have since ever been subject to the obligation to include a statement of reasons explaining the justification and rationale for the provisions established. This obligation, applicable since the entry into force of the Treaty of Rome, and now included in Article 296 TFEU, represents an essential requirement. Failure to include an adequate statement of reasons is a sufficient cause for the Court of Justice to annul the legislation. According to case law, the statement of reasons

> must disclose in a clear and unequivocal fashion the reasoning followed by the Community authority which adopted the measure in question (European Court of Justice 1984).

Moreover, the Court has stated a propos of the proportionality principle that

> if a measure is patently unsuited to the objective which the competent institution seeks to pursue this may affect its legality (European Court of Justice 1987).

Also, the Court may annul an act in case of a manifest error of assessment of the relevant facts.

All these conditions for the validity of an act have in practice involved the requirement for the EU Institutions, the Commission in the first instance, to collect all the relevant information and data and, in the case of risks related to issues with a complex technological or scientific background, to perform an appraisal of the risks in relation to the stated objectives. However, before the introduction of more structured risk and impact assessment practices, such appraisals have been done on an ad hoc basis, in a more or less comprehensive and systematic fashion depending on the nature and complexity of the issue, the policy framework and the public and stakeholder pressure and expectations.

The Court has consistently refrained from substituting its own assessment to that of the legislature and recognised a broad discretion to the EU legislature both in the identification of the relevant data and information and in their appraisal. For instance, repeating on substance some previous statements, the Court has confirmed, in a case concerning the risks of the use of specific metallic additives, in particular methylcyclopen tadienyl manganese tricarbonyl ("MMT") as automotive fuel, that (European Court of Justice 2010):

> … in an area of evolving and complex technology such as that in the case in the main proceedings, the European Union legislature has a broad discretion, in particular as to the assessment of highly complex scientific and technical facts in order to determine the nature and scope of the measures which it adopts, whereas review by the Community judicature has to be limited to verifying whether the exercise of such powers has been vitiated by a manifest error of appraisal or a misuse of powers, or whether the legislature has manifestly exceeded the limits of its discretion. In such a context, the Community judicature cannot substitute its assessment of scientific and technical facts for that of the legislature on which the Treaty has placed that task.

The Court has also observed in relation to amendments introduced by the Parliament and the Council that

> the impact assessment carried out by the Commission, which was annexed to its proposal for a directive… was not binding on either the Council or the Parliament, who, under the co-decision legislative procedure laid down by Article 251 EC, were entitled to make amendments to that proposal.

The Court recalled the duty to take into account scientific and technical data

> Article 174(1) EC provides that the European Union policy on the environment is to contribute to a number of objectives, one of which is the protection of human health. Article 174(3) EC provides that in preparing its policy on the environment, the European Union is obliged to take account of available scientific and technical data. That obligation applies, particularly, to the procedure under Article 95(3) EC, for which it is necessary that new data be taken into account.

The Court recognises a broad but conditioned discretion to the EU Institutions

> Nevertheless, it must be stated that the European Union legislature's broad discretion, which implies limited judicial review of its exercise, applies not only to the nature and scope of the measures to be taken but also, to some extent, to the finding of the basic facts …However, even though such judicial review is of limited scope, it requires that the Community institutions which have adopted the act in question must be able to show before the Court that in adopting the act they actually exercised their discretion, which presupposes the taking into consideration of all the relevant factors and circumstances of the situation the act was intended to regulate.

In the case in question, the Court observing that

> …the Parliamentary Committee also took into consideration various studies challenging the safety of MMT

and that

> The Council maintains that it took into account, during the legislative procedure, studies carried out in 2004 by the International Council on Clean Transportation (ICCT) which demonstrated that the use of MMT is damaging to human health and to the proper functioning of emissions control systems and the Declaration of Brescia on the prevention of the neurotoxicity of metals of 17 and 18 June 2006 which recommends, inter alia, that the addition of organic manganese compounds to fuels should be halted

concluded that

> during the legislative process, the Parliament, the Council and the Commission took into account the available scientific data, including that which emerged in the course of that procedure, in order to exercise their discretion properly. In the light of the scientific documents … the Parliament and the Council did not make a manifest error of assessment in deciding to set a limit for the MMT content of fuel.

Although the particular case mentioned here is of 2010, it reflects a constant doctrine established since older times of EU regulatory action. In light of the above, it can be concluded that a generic obligation on the EU Institutions, subject to judicial review, has applied since the beginning to the EU regulatory policy making, notably to the making of risk regulation, to take into account the relevant scientific and technical data. However, it has been for the Institutions themselves, within the broad margin of discretion recognized by the Court, to establish the criteria, standards and procedures for the proper integration of evidence and science into the EU decision-making process.

Two additional features help understanding the approach that has been applied in the past in order to ensure the quality of certain types of EU regulations. First, in view of the massive body of harmonization legislation planned for establishing the internal market, a Council Resolution of 1985 (European Union Council 1985) launched a "new approach" to technical harmonization and standards. The Communication noting that the Member States have the responsibility of ensuring health and safety requirements in their territories stated that, in order to ensure the free movement of goods, it was necessary to harmonise the relevant provisions, without lowering existing levels of protection in the Member States. Under the new approach, legislative harmonisation was limited to the essential safety requirements (or other requirements in the general interest). The task of drawing up non mandatory technical specifications taking into account the latest stage of technology was entrusted to European organisations competent in industrial standardization (the European Standardisation Committee-CEN, the European Committee for Electric and Electronic Standardisation-CENELEC and the European Telecommunication Standardisation Institute-ETSI). The authorities were required to consider products manufactured in conformity with harmonised standards to conform to the essential requirements established by the Directive. In turn, producers who do not manufacture in conformity with the harmonised standards, have to prove to the national surveillance authorities that their products conform to the essential requirements. Since the essential safety requirements were formulated only in very generic terms, the task to assess specific risks and set design specifications aimed at ensuring the required level of safety was in fact delegated to technical bodies composed of experts from the different interested parties, notably industry. Moreover, the voluntary character of the harmonised standards left a wide scope to manufacturers to choose different specifications provided they could demonstrate their adequacy through risk and technical assessment. The new approach was intended to ensure a maximum level of flexibility, the expert-driven definition of safety specifications and the involvement of interested partied in the setting of standards. It is interesting to note that often the examples of "extravagant" and burdensome EU rules mentioned by the media refer in reality to non-binding standards established by the European standardisation bodies to facilitate the implementation of the very generic "essential requirements" of EU legislation, representing what manufacturers have agreed among themselves!

A second feature of much of the risk-related regulation for the internal market was the obligation on the Commission to evaluate and report to the Council and the Parliament the results achieved with the application of the regulation, problems encountered and need for adaptation. Such reporting requirements provided a framework for ex-post evaluation and adaptation which however was not always implemented in a timely, consistent and substantive fashion. Still, many regulatory changes and up-dating have been introduced following the periodic evaluation foreseen by the relevant legislation. For instance, a General Product Safety Directive (Directive 2001/95/EC) was amended in 2001 following an extensive evaluation of its functioning.

The most important steps toward a more systematic integration of evidence in the EU decision-making process have been taken in response to the food crises at the end of the 1990s and following the European Commission White Paper on European Governance of 2001 that defined the concepts leading to Better Regulation and initiated the process for its development.

7.3 The EU Approach to Better Regulation

7.3.1 The Development, Objectives and Principles of Better Regulation

We have already mentioned Better Regulation in previous Chapters, notably in Chap. 5, in relation to the EU Impact Assessment system, and in Chap. 6, in relation to consultation practices and standards. Here, we wish to discuss Better Regulation as the main framework for the current EU regulatory policy. It should be noted that Better Regulation must be regarded as a process of continuous improvement of EU regulatory practices, which started to develop in a systematic way at the beginning of the first decade of this century, has already involved several steps and developments, is subject to debate and criticisms and is still evolving, with the most recent measures launched by the European Commission in May 2015. We want to briefly retrace here the origin, rationale and development of the EU Better Regulation and discuss its current features and perspectives.

Although the importance of improving the quality of EU regulation had been recognised since the mid of 1980s and addressed in a protocol of the Amsterdam Treaty in 1997, the first really substantive steps to define a more structured and systematic approach aimed at ensuring the quality of regulation were taken in the wake of the Lisbon European Council of 2000, as part of the implementation of the so-called "Lisbon Strategy" setting for the EU the ambitious goal to become "the most competitive and dynamic knowledge-based economy in the world" by 2010 (European Council 2000). It was then noted that high quality and proportionate regulation had an important contribution to make to the achievement of that goal. Following a resolution adopted by the EU Member States' Ministers for Public Administration from EU Member States in November 2000 in Strasbourg, an advisory group of experts was created, chaired by Dieudonné Mandelkern, Honorary Section Chairman within the French Council of State, tasked with developing a coherent approach to improving the quality of regulation and invited to submit proposals to the Ministers, including the definition of a common method for evaluating the quality of regulation. While recognizing the need for regulation and rejecting a simplistic de-regulatory approach (Mandelkern Group 2001).

> Regulation is essential to achieve the aims of public policy in many areas, and better regulation is not about unthinking removal of such regulation,

the Mandelkern Group stated that better regulation should help avoiding excessive regulation (Better regulation "*is about ensuring that regulation is only used when appropriate*"), and ensuring the high quality of the necessary regulation. Better regulation should enhance the credibility of the governance process, prevent the imposition of unnecessary burdens on businesses, citizens and public administrations, avoid damage to firms' competitiveness. Action to improve the quality of regulation was considered necessary for the restoration of confidence in government. The Group concluded that action to ensure high quality regulation, both on the EU and national level, had to be organized around seven core principles: necessity, proportionality, subsidiarity, transparency, accountability, accessibility and simplicity. The Mandelkern Group also recommended a programme including in particular:

- An annual report to the European Parliament and to the European Council on developments in better European regulation by the EU and each Member State;
- Joint training programmes at European level for officials on better regulation;
- The establishment by the Commission of indicators of better regulation;
- Establishment by the Commission of an impact assessment system;
- Commitment by the Council and the Parliament not to consider proposals for regulation that have not been subjected to the agreed impact assessment system;
- Commitment by the Member States to indicate wherever possible the likely broad impacts of significant and substantial amendments they wish to make during negotiation of draft European regulation;
- Member States to introduce national systems of impact assessment for national regulation;
- Adoption by the Commission of a standard minimum consultation period for its proposals of 16 weeks and a Code of Practice for consultation;

- Launch, of a Commission-led systematic, targeted, rolling programme of simplification of existing European regulation in all areas;
- Creation by the Commission of a better regulation network in all regulatory Commission departments, supported centrally as appropriate;
- Drawing up by the Commission, of general guidelines on the use of alternatives to regulation for the pursuit of European policies.

The assessment and recommendations in the early report of the Mandelkern Group have been basically confirmed and further developed in successive documents and programmes of the European Commission leading to the implementation of a structured Better Regulation approach based on three pillars: Regulatory Impact Assessment, Consultation Standards and Practices and simplification and ex-post evaluation of existing regulation (currently the regulatory Fitness and Performance Programme-REFIT). A White Paper on European Governance was adopted in 2001, followed by a Commission Communication on Impact Assessment in 2002, an Inter-institutional Agreements of Better Law-making in 2003, the establishment and updating of Impact Assessment Guidelines in 2005, 2006, 2009 and 2015, the establishment of an internal impact assessment system supervised by an internal board, the establishment of consultation standards as well as a series of successive communications, sectoral guidance documents and reports on various aspects of Better Regulation. A new communication on Better Regulation (European Commission 2015-1) and a proposal for a renewed and reinforced interinstitutional agreement have been presented by the Commission in May 2015 (European Commission 2015-2). That impressive set of documents and the progressive implementation and reinforcement of improved regulatory practices show the importance attributed by the two successive Barroso Commissions to Better Regulation, confirmed and reinforced again by the Juncker's Commission, which has also appointed a Vice-President responsible for Better Regulation with wide powers. Nevertheless, as we will discuss later on, the Commission commitment to better regulation and the action taken have not been without problems and the process to ensure better regulation in the EU is still work in progress.

In order to complete our discussion of the EU risk regulatory policy, we need to make now some considerations on three key aspects: first, what are the reasons and rationale for the commitment to and heavy investment on Better Regulation by the EU Institutions, the Commission in the first instance; second, what policy configuration emerges from the long process aimed at improving the quality of EU regulation, notably in light of the relevant Commission initiatives of May 2015 and, finally, how the EU Better Regulation initiatives have been applied, how they have been received by stakeholders, what problems they have encountered.

There are many theories and models of regulatory policy and in particular on the role and rationale for the regulatory impact assessment (RIA). A discussion on such theories would go beyond the scope and specific subject of this book. We just need to note that the introduction on the level of EU Institutions of Better Regulation and RIA has built on the synergism of distinct but concurrent logics. The first line of reasoning, highlighted since the beginning of EU action to improve its regulatory practice, was the need to restore public and stakeholder confidence in the European project and reinforce the legitimacy of the EU policy and regulatory action. Already the White Paper on European Governance of 2001 (European Commission 2001) highlighted as a major problem to be addressed by governance reform that:

> Many people are losing confidence in a poorly understood and complex system to deliver the policies that they want. The Union is often seen as remote and at the same time too intrusive

and indicated as overall objective of the reform:

> The White Paper proposes opening up the policy-making process to get more people and organisations involved in shaping and delivering EU policy. It promotes greater openness, accountability and responsibility for all those involved. This should help people to see how Member States, by acting together within the Union, are able to tackle their concerns more effectively.

After 15 years, the already mentioned Commission Communication on “Better Regulation for Better Results” of May 2015 reaffirms the same need for change and the objective to restore confidence.

> The European Commission is determined to change both what the European Union (EU) does, and how it does it. The EU, its institutions, and its body of law, are there to serve citizens and businesses who must see this in their daily lives and operations. We must restore their confidence in our ability to deliver.

The second strand of reasons is related to the achievement of economic outcomes. All documents presenting Better Regulation insist on the need to cut “red tape”, improve the environment for competitiveness of EU business and promote the creation of jobs. While the importance of EU regulation is often underlined, notably in relation to the benefits of the internal market and for ensuring a level playing field for business activities across Europe, the accent is on the financial and administrative burden of excessive or inefficient regulation. Although the overall aim declared is sustainable development, to be taken into account notably by including in the impact assessment the economic, social and environmental dimensions in an integrated approach, not surprisingly in a period of economic recession and high unemployment the attention is focused on economic aspects.

Two further important but less visible dimensions of EU Better Regulation should be taken into account in assessing its rationale. One is the wish to ensure a rational policy making where evidence and the scientific input play a greater role. While all EU documents on Better Regulation reaffirm that its processes do not replace the political decision, most of the improved practices are clearly aimed at ensuring better informed decision making and set a more objective frame for decisions. The second dimension, which may play a more or less legitimate and positive role depending on the specific circumstances and aims, is that of political control. Better Regulation has been actively promoted by certain EU Member States and stakeholder groups with the non-declared but easily recognizable aim to put a brake on a EU regulatory activity that they consider excessive and intrusive, by imposing burdensome procedures, delaying action and creating opportunities to interfere with the preparatory work and raise objections even before a proposal could see the light of the day. In a less negative perspective, Better Regulation has been seen as a way to control the tendency of bureaucracy to overregulate. Within the Commission, the accent has been put on coordination, consistency and collaboration between departments, concentration on fewer priorities and careful assessment of the need and added value of regulation, reduction of administrative and financial burden on business and consideration of alternative options. It is worth noting that Member States, while supporting as members of the Council of the European Union the Better Regulation initiatives and the implementation of Regulatory Impact Assessment, mostly as an approach that the Commission should apply to its activities, have shown little appetite for RIA on the national level, in spite of some formal steps in several European countries aimed to set out their own systems, with the exception of the UK and a few other countries that have established well developed and operational national RIA systems.

We have noted above that the EU Better Regulation approach is organised around three main areas: the ex-ante assessment of new regulation, the consultation of and dialogue with the stakeholders and the public and the ex-post evaluation, revision and reconsideration of existing legislation. The current policy of the Commission in each area was summarised in its already mentioned Communication on Better Regulation of 2015.

7.3.2 *Stakeholder and Public Consultation*

Better Regulation sets two interrelated objectives for the stakeholder and public consultation: to ensure transparency and accountability of the preparatory process and to contribute to the effectiveness of the policies and regulation by collecting relevant evidence. Additionally, involvement of the interested parties contributes to better understanding, on various levels, of the EU action. The intention is to cover the entire lifecycle of policies and legislation, from the preparatory work, the discussion and adoption of a decision, until the evaluation of its application.

In order to allow stakeholders to express their views over the entire lifecycle of a policy, Commission's "Roadmaps" and "inception impact assessments" are now open to comments, providing also the opportunity to submit relevant information for the assessment. These are documents made available at an early stage of the preparatory work, when options are still open. Road maps are provided when plans for a new proposal are included in the Commission annual work programme. The "inception" impact assessment is a preliminary version of the impact assessment, still open for change on all its aspects, including the options considered, and not yet validated by the Impact Assessment (IA) Board (see below). The public consultation period is now brought to 12-weeks when preparing new proposals and when evaluating and carrying out "fitness checks" of existing legislation.

Moreover, the Treaties offer the opportunity to national parliaments to provide reasoned opinions on subsidiarity when a proposal is presented by the Commission. The Commission also intends to invite again citizens or stakeholders to provide feedback, within 8 weeks, once it has finalized its proposal, and inform the EU legislator, the European Parliament and the Council, of the comments collected.

The Commission now consults stakeholders also on acts that it has the mandate to adopt under its delegated or implementation powers, setting out technical or specific elements that are needed to implement the legislation adopted by the European Parliament and the Council. It publishes the draft of delegated acts on the Commission's website for 4 weeks, in parallel to the consultation of the experts of the Member States. Stakeholders are allowed to submit comments on important implementing acts which are subject to Committee opinion, during the 4 weeks before any vote by Member States in the relevant Committee.

Under this approach to public and stakeholder consultation, all stages and steps in the lifecycle of EU policies and regulations of all kinds, including both general and implementing risk measures, are open to external scrutiny and input, with a timing that allows the taking into account in the decision making process of the feedback and information received.

Two additional measures are aimed at ensuring effective communication with the public and stakeholders. The already mentioned "explanatory memorandums" that accompany Commission proposals will focus on how better regulation principles have been applied, notably the reasons for the initiative, for the tool used, the input provided by stakeholders and what are the likely environmental, social and economic impacts. They will also include a subsidiarity and proportionality assessment of the initiative.

Finally stakeholders are enabled to provide feedback on any aspect of EU policy and legislation, at any time. The objective of the Commission is to monitor how EU regulation affects people and businesses, in view of possible improvement. This objective is implemented by a new feature on the Commission's better regulation website, called "Lighten the Load – Have Your Say". Input received is followed up directly by the Commission, or submitted for further consideration to a new stakeholder platform (see below).

7.3.3 Ex- Ante Impact Assessment and Ex-post Evaluation

Impact Assessment The Impact Assessment (IA) approach is based on integrated consideration of economic, social and environmental impacts. As we have already seen in Chap. 5, the assessment of risk-related regulation must take into account results of formal risk assessment made by EU specialized bodies. Reference is made in the Commission's IA Guidelines to the commitment to use the best available evidence and science. A new stress is put on the need to monitor and assess existing plans and measures and to factor the results in the assessment of new initiatives. Although the special needs of SMEs are recognized, it is highlighted that when it comes to health and safety requirements, SMEs could not be exempt based on economic considerations from the necessary protection rules. A new IA Board is established (European Commission 2015-3), replacing the one set in 2006, charged with assessing the quality of the Impact Assessment, which can now exert a considerable power because of the obligation imposed by the Better Regulation rules on the Commission departments to obtain the approval by the Board of their draft IA before a proposal may appear on the agenda of the Commission for adoption.

Following insistent stakeholder criticism on the composition of the IA Board and a request from the Parliament for an independent Board, the new Board, which is now called "Regulatory Screening Board" is composed of a Chairperson plus six members working full time for the Board with no links to Commission departments, with three of them recruited for the task from outside the Institution for fixed non-renewable terms. The Board will also assess the quality of the "fitness checks" of existing legislation.

The Ex-post Evaluation of Policies and Regulation Following a number of ad hoc initiatives for the evaluation of existing legislation launched since the 1980s, a Regulatory Fitness and Performance Programme (REFIT) was launched in 2012. The objective of the Commission was

> ensuring that EU legislation remains fit for purpose and delivers the results intended by EU law makers.

Once again, the Commission states that (European Commission 2015-4)

> REFIT is not about deregulation but rather about regulating better.

Although a comprehensive, integrated evaluation approach, based on the same principles of the Impact Assessment is announced, the focus is clearly put on economic aspects: removing red tape and lowering costs without compromising policy objectives. The important feature of REFIT is its permanent character and a "*lasting commitment to keeping the body of EU law lean and healthy*". In implementing REFIT, the Commission intends to target the legislative instruments and the specific aspects that may be a source of significant inefficiencies, to perform quantitative evaluation of costs and benefits and to collect feedback and suggestions through a REFIT Platform composed of experts from business, social partners, and civil society, experts from the 28 Member States, the European Economic and Social Committee and the Committee of the Regions (European Commission 2015-5). The Platform, which is chaired by the First Vice-President of the Commission, may provide its own suggestions on how the legislation can be improved. It is also in charge of examining concerns or suggestions presented by stakeholders. The Commission is committed to react to and explain how it intends to follow up the Platform's suggestions.

The REFIT Programme is aimed at supporting the continuous adaptation of EU legislation, in light of the results achieved, the effectiveness and efficiency of the measure evaluated, the problems encountered or created by its implementation, the new needs and the changing economic, social, natural and environmental context. Action may be needed to adapt, simplify or simply withdraw certain policies or regulatory instruments.

Finally, a very important new feature of REFIT is that, differently from previous initiatives, the scope of the evaluation may now be extended to cover whole policy areas. An example is the comprehensive evaluation of the EU food safety legislation performed by the Commission.

7.3.4 *Issues and Positions on Better Regulation and Its Implementation*

The current configuration of the EU regulatory policy, as updated by the Commission initiatives and proposals of 2015, is characterised by a greater recognition of the need to involve in implementing Better Regulation all the layers of the complex EU decision making process, as well as the Member States for their implementation responsibilities. The Commission has proposed a new Interinstitutional Agreement, which includes some far reaching commitments. Particularly relevant are:

- To prioritise simplification and improvement initiatives, ensuring their rapid adoption;
- To carry out a comprehensive impact assessment on any substantial amendments that the European Parliament or the Council propose during the legislative process, before any final decision is taken. Each institution may call for an independent panel composed of independent experts, appointed by each Institution, to carry out an assessment of any substantial amendment to the Commission proposal;
- To include in new acts provisions for the monitoring of their functioning, and for their future evaluation, and examine whether legislation already in force may achieve the desired objectives, before considering new initiatives;
- To request Member States to avoid "gold plating" of EU rules or to explain the reasons for gold-plating when it is considered necessary.

Since its inception, Better Regulation has been welcome in principle by the EU Institutions and the Member States, by part of the stakeholders, notably business, by trade partners and international organisations, notably the OECD, which is active in promoting regulatory improvement and the diffusion of RIA among its members, and by part of the observers and scholars. It has however been regarded with suspicion by most of the NGOs and other organisations with an interest in health, safety, environmental, consumer and worker protection, which have seen it has a business-lobby driven approach aimed at deregulation, the lowering of European protection standards, compression of democratic control and excessive influence of vested interests over public policies. On the EU level, the European Commission has consistently insisted on a balanced approach, rejecting the allegation that Better Regulation would allow short term and unilateral business interests to prevail over other public interest imperatives or over political choice and democratic accountability. Nevertheless, in a period of prolonged and structural economic crisis and high unemployment in most of the EU, it would be naïve not to recognise a tension, within the short-term horizon of politics, between on the one hand the pressing needs for competitiveness and growth promotion and, on the other hand, maintaining and reinforcing protection standards.

In fact, Better Regulation, in spite of a generic recognition by all the stakeholders of the need for improving the quality of EU regulation, has been criticised under different perspectives and for different reasons. While certain stakeholders consider it to be part of a deregulatory policy, the supporters of its principle have criticised the way in which it is implemented in the EU context, claiming that it has not shown its potential because of political resistance and inadequate methods, tools, scope, resources and procedures. As we will see, the recent Commission initiatives try to address the most relevant criticisms and the difficulties encountered.

The introduction of Better Regulation with its RIA pillar has been a slow process, with mixed results so far, which is still in progress, not surprisingly account taken of its ambitious objectives, the wide and diverse scope of application across sectors and the complex EU decision-making context. The European Commission has monitored the development of the system and taken successive steps to complete and reinforce it, as explained above. Performing integrated impact assessment has been a serious challenge, representing a significant burden in terms of time and resources. In the first year of application of IA, only 43 out of 580 proposals could be submitted to IA. A report commissioned by the Commission, published in 2007 (The Evaluation Partnership 2007) on the "*Evaluation of the Commission's Impact Assessment System*" showed that RIA was received with mixed feelings in its first years and encountered significant initial difficulties. Such assessment was based on opinions and perceptions of the Commission officials and of Members and staff of the European Parliament, as well as on the examination of a sample of the impact assessments reports prepared by Commission departments. In fact there was initially some scepticism on the possibility to identify and quantify all the relevant impacts in an integrated perspective. Quantification of social and environmental impacts was considered particularly problematic. The examination of the impact assessments performed showed a poor quality in many cases, with a more developed assessment of only the preferred option and an understandable tendency of the leading Commission department in charge of the impact assessment to focus on the aspects and dimensions directly relevant for the policy under their responsibility. Nevertheless, it appeared that the IA process was an effective framework for stimulating collaboration and co-ordination between the Commission departments thanks to the systematic establishment of inter-service groups contributing to the IA and the supervision of the General Secretariat of the Commission.

Since the early days of IA, the reviews by the IA Board have become increasingly stringent and critical overtime. Nevertheless, at the time of the assessment study mentioned above, there was some scepticism on the usefulness of IA results for the EU decision making process due to several factors notably the type and timing of the information produced, not always answering decision makers' questions and not always available when the information could have a significant influence on decisions, and the difficulty to reconsider already established policy orientations and to overcome based on a technical assessment, political and legal constraints.

Although the implementation of Better Regulation and its RIA have actually improved openness and communication practices, its practical implementation has been criticised on several points. As said above, certain NGOs, as well as part of the Members of the Parliament and some academic observers consider that the system is unbalanced, serving mostly business interests and a de-regulatory strategy. As a researcher as noted (Redaelli 2005)

> if actors think that RIA is tilted towards one actor's preferences, there is no economic analysis that can compensate for the credibility deficit.

The level of mistrust on Better Regulation by part of NGOs and civil society organisations is well represented by the initiative involving more than 50 trade unions and NGOs, launched in May 2015, to establish a "Better Regulation Watchdog" (http://www.betterregwatch.eu/) with the aim to watch the development and functioning of Better Regulation in view notably

> to increase understanding among civil society, media and decision makers about the risks and potential negative impacts of the Better Regulation agenda on social, environmental, labour, consumer and public health standards for citizens and the public interest.

On a practical ground, the period allowed for replying to consultations has been considered too short for preparing positions on complex issues. NGOs have lamented not to have sufficient resources to compete with business in a technical and scientific debate. Business organisations have in turn claimed that often IA is focused on justifying pre-defined political decisions. The quality of certain IA has been criticised and the IA Board has not been considered sufficiently independent from the Commission departments producing the assessment and from the political masters of the Institution. There has been scepticism on the quality and quantity of the resources available for the assessment. While observers have recognised the efforts made to consult and to show openness to stakeholder views, the consultation practices were criticised by part of the stakeholders for not offering the opportunity to comment on the impact assessment before decisions were taken by the Commission, and for not allowing stakeholder to provide their input over the entire life cycle of policies and regulation.

It should again be noted that the initiatives taken by the Commission in 2015 try to address most of the criticisms and difficulties encountered. Nevertheless, in spite of the efforts to dispel concerns on deregulatory, business-tilted intentions, civil society organisations represented by the above mentioned Better Regulation Watchdog have expressed a critical view also on the 2015 initiatives, notably on the new interinstitutional agreement, raising the attention of the European Parliament to the risk that

> the proposal as drafted would, in our view, considerably hamper the ability of the European Parliament to use its full post-Lisbon powers as a co-legislator.

In a letter to the President of the European Parliament (http://www.betterregwatch.eu/), the Watchdog organization raises five issues that they urge the European Parliament to address

> 1. The Better Regulation initiative should not threaten the democratic powers of the European Parliament and the Council
> 2. The Commission should not propose outright blanket exemptions of EU legislation
> 3. Impact assessments should be used to improve, not threaten legitimate regulation
> 4. Transparency should not lead to even more corporate influence on law-making
> 5. Better Regulation should not lead to regulatory convergence to the bottom.

The Better Regulation approach poses some substantive problems that can only be solved by clarifying its objectives and its role in the decision-making process, moderating expectations, ensuring a really balanced and credible approach, balancing the resources invested and the added value of the process in a pragmatic and realistic fashion.

A necessary condition for achieving the objectives of the EU Better Regulation is the full and active involvement of the European Parliament, the Council and the Member States. The Commission has taken the initiative of more comprehensive inter-institutional agreement in that respect (European Commission 2015-2). However, it is difficult to imagine that any such agreement could succeed in practice without a consensual clarification of the role of Better Regulation, notably of its impact assessment system.

We have already discussed the complexity of the EU decision-making process. It reflects the unique nature of the EU and its internal articulation. Behind the complexity of the EU process itself, there is also the variety of interests, situations, priorities, political systems and circumstances, historical heritages, values and public demands in an ensemble composed of 28 sovereign Member States, with more than 500 million citizens. Agreement on common policies and rules on issues related to health safety and environmental risks, often charged with emotions and seen through the filter of diverse values, perceptions and concerns, can only be reached on the EU level through political negotiation, bargaining and search for balanced compromises. Value-based judgement and political-balancing of interests play necessarily a major role in that respect. Once again we have to insist that science, evidence, impacts assessment, including in particular risk analysis, and the formal procedures of Better Regulation can only help by informing and framing the process, notably the participatory approach discussed later in the Chapter. But the researcher and expert perspective cannot replace the political appraisal and negotiation, particularly under the EU circumstances. Overstating the potential of IA to determine decisions, creates excessive expectations and inevitable frustration. And the political decision makers would be more willing to invest on consideration of technical assessments if it were clear and accepted by all that the role of experts in the decision-making process is to better inform, not to constrain the political process, or to transfer to scientists and "bureaucrats" a significant part of the responsibility for policy and regulatory choices.

At the time of the Commission new initiatives in 2015, Better Regulation has significantly contributed to the internal inter-service co-ordination, to the improvement of the wider consultation practices, to the quality of the technical assessment of policy options. Nevertheless, Better Regulation has clearly not yet achieved its objectives to restore citizens' confidence and improve the public perception of the legitimacy, quality and added value of EU regulation. The traditionally controversial risk issues like GMOs, endocrine disrupting substances, nuclear power or oil fracking have not become less controversial. It is difficult and possibly still premature to try to assess to what extent the added value of Better Regulation in relation to regulatory quality improvement, higher welfare of citizens and better conditions for business and stakeholders is materialising and whether the actual gains over the "business as usual" baseline exceed the additional administrative costs. The new Commission initiatives represent a substantial challenge. They require significant resources for the effective implementation of assessment evaluation and consultation commitments. They involve a huge administrative and technical effort and adds heavy procedural requirements to the already complex and slow EU decision-making process. Their success depends also on the regain of confidence of all sides of the civil society, achieved through a balanced implementation approach, not focused unilaterally on the cost-side and not seen as unduly influenced by business interests. Success of the EU IA also depends on further development and balanced implementation of methods that correctly assess and represent costs and both monetary and non-monetary benefits, on a proactive involvement of all layers of actors in the regulatory cycle, based on principles and working procedures both practicable and respectful of the role and nature of the political decision makers, and last but not least, on clarity and honesty about the limitations of impact assessment, and recognition of the legitimacy of negotiated political decisions in the EU democratic institutional configuration. Overall, the structure, content and implementation of Better Regulation and RIA need a continuous improvement process. They have been introduced and shaped so far mostly on the basis of political considerations, generic aims and pressure from economic needs and business requests. There is still a need to gain experience and collect evidence in order to determine and demonstrate the configuration, methods and level of effort producing actual net benefits in terms of welfare, with an overall added value exceeding the costs. Although the Impact Assessment procedure is a powerful tool to improve the quality of regulation when used in a neutral, timely and proportionate manner, it may become a source of abuse and inefficiency when used as a political control tool or applied in an inappropriate way. A worrying example is provided by the already mentioned (in Chap. 4) impact assessment of the "scientific" criteria for endocrine disrupting substances. Although it is perfectly reasonable to explore the impact of a set of criteria that may have huge consequences on the availability of certain products and the future of important industrial sectors, it is completely wrong to give the impression that the choice of scientific criteria could be based on the results of an impact assessment. Nobody would propose to make an impact assessment of the Newton laws or the Second Principle of Thermodynamics in order to adjust them in case of inconvenient consequences! Of course, the legislation itself might need to be re-modulated in light of the results of an impact assessment, but the scientific criteria in question may only be set in light of scientific knowledge on the effects of the substances considered on the endocrine system. The Court in its conclusions on the case brought by Sweden, other Member States, the European Parliament and the Council against the Commission for the delay in the approval of scientific criteria for identifying endocrine disrupting substances (Court of Justice 2015) has stated that there were no provision of the biocidal products regulation requiring an impact assessment and therefore the "need" for such an assessment could not be considered a justification for the delay in the adopting the criteria. This conclusion shows the importance of an appropriate sequence and timing of the various assessment steps in preparing decisions. The impact assessment remains a technical decision-aid tool, not an essential procedural requirement recognised as such by the Court.

Finally, it cannot be ignored that an excessive focus on simpler, flexible and more lenient regulation may result in an outcome opposite to that hoped, with high cost to business, market disruption, loss of competitiveness and jobs loss, as the Volkswagen case, which has been made possible by the inadequate, easy to circumvent test requirements on pollutant emissions from motor vehicles, has shown. On the current development stage of the impact assessment methods, the business, economic and social risks of more flexible but less stringent and possibly less effective regulatory options are simply not factored in.

7.4 The Role of Science and Evidence in the EU Regulatory Policy

7.4.1 The Role of Science in the Regulatory Decision Making

It is very often claimed that policy and regulation, and in particular risk regulation, should be science and evidence-based. And this book takes firmly the stance of science as a key tool for informing and supporting decision making and for providing the means of effective risk management. But, what is science? How to distinguish sound science from pseudo-science? Is there a science of risk? Is risk assessment "science"?

There are many definitions of science. Most frequently they refer to knowledge acquired through the systematic study of the structure and functioning of the physical world, doing experiments and measurement to confirm the validity of hypotheses, and developing theories allowing linking, explaining and anticipating experimental results. The modern scientific method has characterized natural science since the seventeenth century and is based on systematic observation, measurement, and experiment, and the formulation, testing, and modification of hypotheses. In turn, propositions resulting from the application of a system of thought or a theory that do not comply with the generally accepted scientific standards are considered "pseudo-science".

There are also many theories of science. The most useful and pragmatic features to recognise science are those related to the process through which it is produced. The starting point is normally the observation of empirical facts followed by the formulation of hypotheses and modelling of the relationships between such facts. Such hypotheses and the validity of the models must be testable and falsifiable. In fact, the scientific method is characterised by the verification through controlled experiments or observations of the validity of hypotheses in explaining, representing and anticipating empirical data on the behaviour of observed systems. In the case of natural sciences, theories are accepted (at least provisionally) if experiments and observations are feasible in order to confirm or "falsify" the hypotheses made and provide results consistent with the hypotheses. The tests must be adequate to exclude that a positive results be due to chance and coincidence. A particular importance is given to controlled experiments that can discriminate against the influence of confounding factors. Formal requirements also apply such as documentation, repeatability, openness, transparency, peer review and absence of bias. Also arcane theories like Einstein's General Relativity and quantum mechanics have been subject to and confirmed through direct experimental verification of their implications. Is this type of verification feasible in the case of risks? In most cases, only indirectly, partially and with some degree of uncertainty. The toxicity of chemicals is tested on animals and then extrapolated to humans based on hypotheses that are not directly verifiable. Risks of nuclear accidents can only be "guessed" through probabilistic methods and there is no way to confirm the results ex-ante with direct, controlled experiments on entire complex engineering systems. The hypothetical long-term effects of radio-frequency electromagnetic fields for which currently no plausible mechanism of action is known could possibly only be observed when the risks materialise and become in fact actual health damage. In Chap. 4 we have introduced in detail the methods of risk assessment in various sectors, showing how they depend on (reasoned) assumptions, extrapolations, defaults, professional judgment, expert opinion elicitation etc., all attempts to approximate the "real world" for which often there are no directly relevant, specific empirical data. The disciplines that intervene in risk assessment, for example animal toxicology, pharmacology, chemistry etc., are indeed branches of natural sciences that comply with the substantive and formal requirements of the scientific method. But, we re-state here that risk assessment is not per se "science". It is a systematic and rational process that makes use of scientific disciplines, filling the inevitable gaps with conventional choices and ad hoc pragmatic approaches. The debate on the role of science in risk regulation is often confused by misunderstandings on the nature of risk assessment, due to the lack of background in both natural sciences and risk assessment of many of the scholars and social scientists researching the subject. Moreover, although the difference between science and pseudo-science is easily recognisable in the case of natural sciences, the distinction may be less clear-cut when the mainstream science is not strictly based on the scientific methods and it validity depends at least partly on convention, consensus and recognition, notably the "status" of the body "pronouncing" the science. In some cases, it would be proper to talk of mainstream and minority opinions, being clear that "pure" science itself is full of examples of minority opinions being finally recognised as being "true".

With these considerations in mind we can better see now what science, as used in risk assessment, can deliver in support to decision making and what instead should not be expected of it in complex cases.

Excessive expectations are:

- A clear statement on risk or safety
- A rapid and final response even when specific data are not available
- A complete and final consensus
- Certainty on future outcomes
- Lifting political responsibility
- Preventing the need for precautionary measures or the risk of future unexpected losses.

But the systematic approach of science-based risk assessment has a tremendous potential for:

- Framing rationally and possibly quantitatively the debate and the decision-making process
- Representing the hard facts and identifying the knowledge gaps and uncertainties
- Indicating the perspectives, priorities and timescales for knowledge improvement

- Eliminating pseudo-risks or at least exposing the purely hypothetical nature of certain alleged risks, clearing the way for identifying the mechanisms and reasons for risk perception and concerns
- Providing a basis for comparing risks on a consistent approach.

In light of what we have said on the nature of science-based risk assessment, the best use and the real added value of "science" in the regulatory process is to help structuring the analysis and the appraisal of risks, and informing the discussion, deliberation and decision process. Science becomes suspect for part of the stakeholders and policy makers when it is presented as a substitute for a broader "political" mediation between the facts, uncertainties, interests and values that most often intervene in dealing with controversial risk issues. Science is in fact sometimes presented as (DeLeon, Dryzek and Bobrow, Majone, Stone, cited in US NRC 2012) "a means to rescue policy from short-sighted influence peddling and power politics".

In this perspective, the public may see the decisive role that would be implicitly or explicitly attributed to scientific experts as a way to deprive them of the right to have a say, through accountable political institutions, on their health, safety and the environment. Such concerns are confirmed by the interpretation of the role of science which is proposed by some scholars (Majone 1996) considering that "reliance upon qualities such as expertise, credibility, fairness, or independence has always been considered more important than reliance upon direct political accountability".

According to these orientations politics should be replaced by "rational" decision making, experts should be empowered to decide and decisions should be, according to the same Author, "protected from political interference, patronage politics and emotions of the public".

The underlining assumption is that "science", science alone, is in a position to determine the best way to address and manage risk issues.

In order to highlight the appropriate role for science-based tools in the EU regulatory process, under the EU conditions, it is therefore necessary to discriminate between different aims and motivations behind a rationale for science:

- <u>The ideologically motivated rationale</u>: science as a means to ensure less state (and therefore less regulation), less Europe, less public intervention, within a de-regulatory perspective. As we show in other parts of the book, the hidden assumption here is that since risk science is affected by uncertainties and dependent on data that most often is in the hands of those actors who have an interest in avoiding regulation, requiring full scientific proof and excluding political mediation would lead to "laissez faire" in many cases, or re-nationalisation with a resulting competition between regulatory systems toward the less demanding protection level.
- <u>The vested-interest driven rationale</u>: "we know better"; let experts and industry decide the best way to address risks; less politics, less emotions: here the point is not to de-regulate for ideological reasons, but to claim that the need and content of regulation should be better determined by experts, based on scientific and technical assessment that they alone have the competence to perform and interpret.

- The effectiveness/efficiency oriented rationale: this approach recognises the relevance and the legitimacy of societal concerns, values and expectations, the need and legitimacy of a decisive role for the political institutions in our democratic systems and the limitations of science in assessing risks. Nevertheless, it claims that full consideration of science-based assessment and specific, relevant evidence is necessary in order to determine the most effective and efficient way to achieve the public-interest objectives that strike a viable balance between the aims, interests, concerns and impacts at stake. This is the rationale for science in decision making that is consistent with the European Union principles of action and decision-making procedures (and is supported by this book, as the reader has certainly noticed!).

An idealised model of rational decision making would require:

- A clear identification, definition and characterisation of the (risk) issue to be addressed, based on specific evidence and data and on quantitative assessment;
- The setting of shared and generally recognised goals, values, criteria, metrics;
- The identification of a complete set of suitable alternative options with all the descriptive parameters needed for characterising and comparing them;
- The comparative evaluation of the options based on (quantitative) determination of their results, consequences and impacts, with a multi-criteria approach allowing comparative consideration and weighing with appropriate metrics against the set goals, values and criteria;
- A consensus-based determination of the "best" option, ensuring the best perspectives for the effective and efficient implementation of the decision, including the need for ex-post evaluation and possible adaptation in light of results and new knowledge.

It is clear that this idealised approach is based on conditions, notably the availability of data, scientific knowledge and methods, but also consensual goals, values and criteria, that are rarely met. The current practices try to approximate such ideal conditions, notably through participatory governance and improvement of scientific risk assessment methods, processes and procedures, as explained later on in this Chapter and in Chaps. 8 and 9.Evidence and the information provided by risk assessment are also relevant and recognised in order to enforce the accountability of institutional decision makers, both protecting legitimate interests against "capricious and arbitrary" decisions and, in the opposite direction, ensuring that significant indications of potential risks are not ignored for political expediency or complacency toward vested interests.

7.4.2 Evidence Based Risk Regulation

It is common sense that evidence is of the utmost importance in order to identify and characterise the risk issues to be addressed, to identify and compare options for action and finally make an *informed* choice. But things become more complicated when it is claimed that policy and regulatory decisions, in order to be effective and efficient, must be *based* on evidence. What means *based*? And what is effective/ efficient decision making? And, finally, what evidence is relevant?

Claims that risk regulation should be "evidence-based" sometimes fail to recognise the complexity and multidimensional nature of the political decisions. Politicians, who in the end have to assume the responsibility of decisions, have their own perspective of what is "effective". They need to take into account divergent interests and the weight of public expectations, perceptions and values. They want to be re-elected, need to respond to their constituencies, and often also to seek the approval and support of certain sectoral interests. Moreover, the political arena is not a consensual place where the main aim of the actors is primarily to build consensus based on evidence. On the contrary, depending on the subject and the political background, politicians may tend to give prominence and visibility to what divides. On nuclear safety for instance, the Greens would not like to be seen in the same boat with pro-business right-wing parties, whatever line of reasoning "evidence" would support.

In fact, science and policy are not necessarily convergent activities. Science, and science-based approaches like risk assessment, tend to determine results that can be endorsed by all those applying the scientific method. Political decision making acts in the realm of divergent interests, ideology, competition between political parties. Contrary to what is sometimes said in the media, referring to an ongoing technocratic drift, our democracies and the EU are based on a *decisionist*, not a *technocratic* model. The last word on decisions is with the political decision makers, not the experts. Often, those criticising a "democratic gap" of the EU decision making, are in fact concerned not about a lack of political accountability of the EU Institutions, but because of the little responsiveness of the Institutions on a particular subject to *their* particular interests and desiderata.

All that being noted, we conclude, with a metaphor, that regulatory decisions on risk could <u>not</u> be expected to be taken by computers even if and when it were possible to collect all the necessary evidence and data that can be provided by natural and social sciences, and to include in representative and validated models weights representing the market and non-market costs and values, the societal preferences, perceptions and concerns. Evidence-based decision making could not be interpreted as collecting, weighing and elaborating data and evidence, and drawing in a mechanistic way unchallengeable conclusions that lead to undiscussable decisions in a linear sequence.

Should we conclude then that it is hopeless to ensure a role for evidence in the decision making process? Of course no. All this book is based on the opposite assumption. But we wish to warn the reader against a simplistic vision of risk-related decision making. Once again, evidence must inform and frame decisions. It should prevent *arbitrary and capricious* decisions. It makes politicians accountable for their decisions against a background of facts illuminated by the use of scientific tools. Then, other dimensions also come into play. Risk analysis is part of a complex, dialectical process. The expression "evidence-based" decision making should therefore be used with some qualification. The first condition to make science useful for risk decision is to be aware of and accept the world of decision making as it is.

7.4.3 Adaptive Risk Regulation

We have shown in Chap. 4 that risk assessment is most often affected by uncertainties. Regulating new technologies, products or developments poses particular problems since data and scientific knowledge on their potential risks often become available only when they are already in the market place since long. Sometimes, evidence on adverse health or environmental effects becomes compelling only based on epidemiological studies or observations on ecosystems, years after the introduction of the responsible technology, when substantial damage is done. And once a technology is widespread, banning or restricting it may involve huge economic and social costs and disrupt the functioning of entire technological systems. As will be discussed in Sect. 7.5, there is plenty of examples of cases where late and costly remedial action has had to be taken: asbestos, CFCs, lead in petrol, PCBs, BSE etc. Rational and efficient regulation should try to anticipate risks, notably based on the Precautionary Principle. But, the risk is also high to over-regulate, putting a brake on innovation, for the same reasons mentioned above: the lack of sufficient data and adequate scientific knowledge of the potential adverse effects at the early stages of development and deployment of a new technology. While in some cases precaution may justify a stringent preventive approach (notably: no data, no market), the needs of innovation can often be better addressed through an adaptive approach. The first steps may focus on knowledge improvement and preventing possible irreversible damage. Adaptive regulation involves continuous monitoring of effects, collection of data and parallel research on plausible potential risks. The initial constraints on the development and use of the technology or products considered should be proportional and limited to the objective to mitigate the potential risks and to prevent serious, irreversible damage, should the concerns materialise, and be timely reinforced or waived, as appropriate, as soon as new evidence becomes available. Adaptive regulation requires a cultural change both on the side of regulators and by industry and promoters. It can work well if all the stakeholders collaborate constructively.

7.5 Risk Regulation, Innovation and Competitiveness

7.5.1 Innovation and Risk

Innovation leading to new products, processes, technologies, services and business or activities and behaviour is by definition related to the concepts of newness and change. And something new is likely to be less well known of the older thing. At the beginning of a new development or technology, there is no or little experience on its working and its consequences. Of course, if an innovation has promoters, has been backed by investors, has a market, it must be assumed that it brings certain benefits. But on the side of the possible negative consequences, the associated hazards and their probability to materialise, the "unknown unknowns" prevail. Not always however. There have been innovations, for example in the financial sector, whose risks were identifiable since the beginning but have been overlooked. Nevertheless, in general there is a time gap between the introduction of an innovation and the availability of the data, science and information needed for a risk assessment. There is also, very often, an asymmetry of knowledge between developers/promoters of the innovation and risk assessors and control authorities. Only the organisation developing an innovation has good knowledge of it at the beginning. The developer has no interest in disclosing more than what is required by law, in order to both protect its property rights and not to expose aspects that could raise concerns on potential risks.

The difficulty of innovation control was well expressed by David Collingridge of the Technology Policy Unit at the University of Aston, already in 1980, in his book *The Social Control of Technology* (Colligridge 1980), with the so-called *Collingridge dilemma* stating that impacts of new technologies cannot be easily predicted, for the reasons mentioned above, until the technology is extensively developed and widely used, but control or change of the technology becomes costly, painful and may be disruptive when the technology has become entrenched. Collingridge put it simply, stating in synthesis that it is impossible to foresee the need for change at the early stage of a development, when change is still easy, while when the need for change becomes apparent, changing course is expensive, difficult and time-consuming.

As an example of the difficulty to foresee impacts of new technology, Collingridge recalled that the UK Royal Commission on motor vehicles in 1908 concluded that the most relevant problem of the introduction of motor vehicles was likely to be the dust from automobile traffic in non-surfaced roads!

Nowadays innovation is among the magic words of politics. All economic and social problems will be solved only through innovation, our economies and our companies will return to competitiveness thanks to innovation, etc. Once again, oversimplification here leads to a discourse which is less and less credible for the public. Everybody knows that innovation may be bad. It has often be bad. It can cause the disruption of traditional activities. It can create toxic financial products. It has dramatically increased the destructive potential of weapons. Innovating is not necessarily a win-win development. Often there are winners and losers and what is gained by winners may be less than what is lost by the losers. There is plenty of books and articles on the subject. Was the innovation based on financial products incorporating sub-prime loans good? Some made a lot of money from them, but what about the costs, losses and societal disruption generated across the world by the subprime crisis? Was asbestos a good innovation? And CFCs?

Innovation may indeed be accompanied by serious risks. We have already discussed how risk-taking is essential for progress. But rational risk-taking, using all the tools of science to prevent or mitigate risk as far as possible, is different from crazy adventure, jumping without a parachute. There is a need to incorporate the risk analysis paradigm into innovation development at an early stage, with the appropriate instruments and resources. Development of innovation and production of data and scientific knowledge on its possible adverse effects should proceed in parallel. This is what much of the EU risk regulation is about. Some see it as a burden, penalizing the EU economy and competitiveness. Where developed and applied in an efficient and proportionate manner, it is the opposite: it reinforces the economy by promoting sustainability and providing an insurance against bad and costly surprises.

7.5.2 Competitiveness and Risk

Competitiveness is another of the magic words of today's politics. Growth and jobs depend on competitiveness. In fact, the repetition and apparent simplicity of statements like this hide the great complexity of the relationships between growth, employment and competitiveness. On the EU level, the background is however provided by the Treaties' requirement on the sustainable development of Europe, integrating economic, social and environmental sustainability. Therefore, competitiveness should not be pursued by sacrificing environmental and social goods, citizens and consumers health and safety. Reducing production costs by increasing the associated environmental or health costs, imposed on society in the form of negative "externalities", is not a sustainable "competitive" regulatory strategy. Instead, it should be considered competitive to switch production to more sophisticated, less polluting and more efficient technologies. The EU has always maintained that the search for competitiveness should not lead to a reduction of health and environmental standards in its regulation, and must instead be framed within a broader sustainability strategy.

In conclusion, it would be wrong to set a hierarchical relationship between the needs of innovation and competitiveness and the protection against health and environmental risks. Decision makers should aim to establish a balance between these aspects with the support of the tools provided by risk analysis and better regulation.

7.6 Risk Governance

7.6.1 The Concept of Governance in Risk Policy

As we have observed several times, not all risks are subject to public policies and regulatory action. Our life is full of risks and dealing with risks is in the first instance an individual responsibility. We have discussed the reasons, justifications and domains of public intervention aimed at risk prevention or reduction. And we have seen that the EU only acts on risks where and as far as the Treaties confer upon it a specific competence, the objectives of the action cannot be sufficiently achieved at national level and the content and form of the action does not exceed what is necessary to achieve those objectives. When such conditions are met, the Treaties attribute to the EU institutions the responsibility of establishing the appropriate policies and measures, according to clearly defined rules and procedures. Overall, the Treaties establish a legal, institutional and procedural framework for "governing" certain categories of risks.

While the legal and political responsibilities of the EU institutions as well as the EU decision making process are defined by the Treaties, and could be modified only through Treaty changes, the EU policy making practices have flexibly evolved over the years towards a more open and participatory approach, in view of experience, with the aim to meet new public expectations. Such an evolution, which has taken place across the various EU policy sectors, in parallel to the increasing role of the EU on health, safety and environmental matters, has a strong relevance in shaping EU regulatory practices, notably in the risk management area. The current EU decision making principles and practices recognise and ensure a greater role for the civil society. In that respect, several requirements of the Treaties themselves are relevant, notably the first three paragraphs of Art. 11 of the TEU:

TEU Article 11

"1. The institutions shall, by appropriate means, give citizens and representative associations the opportunity to make known and publicly exchange their views in all areas of Union action.

2. The institutions shall maintain an open, transparent and regular dialogue with representative associations and civil society.

3. The European Commission shall carry out broad consultations with parties concerned in order to ensure that the Union's actions are coherent and transparent.

…."

Because risks most often involve both "technical" and "societal" aspects, openness and participation by the civil society are important for preparing effective management decisions. We have discussed in some detail the concept of risk as a "social construct" and its "contextual" dimensions as well as the relevance of perceptions in shaping and framing the societal dimensions of risk. Since risks are about possible future, uncertain events and outcomes, they take form in our mind before possibly appearing as actual physical or economic outcomes. Effective risk policies and measures must address, take into account and balance the interests, perceptions and values of the interested parties and of the wider public. For complex and systemic risks, the success of the intended measures, notably their effective implementation, depends on the engagement, participation, contribution and action of a variety of actors. Moreover, the EU institutional decision-making process does not take place in an empty space. In a democratic society, and particularly in the complex and diverse EU context, decisions are influenced by a variety of stakeholders, actors, interests, which are articulated in multiple and diverse layers: regional, national, international, sectorial, professional etc. Therefore, the risk analysis paradigm needs to be framed in such wider and complex societal context. To that aim, it is necessary to develop and organise a wider participatory framework integrating, in an ordered, balanced and transparent way, the civil society's contribution with the EU institutional decision making process. The concept of "government" must be broadened towards that of "governance". That being said, we need to warn at the very beginning of this discussion on governance that while an open, participatory approach is a "must" for ensuring better decision making in case of risks which are characterized by complexity or are controversial, notably because of the involved uncertainty or ambiguity, such an endeavour is not without its own difficulties, limitations and risks, which will be briefly discussed further on in this Chapter.

Governance is a relatively recent term in the EU political and institutional lexicon. The term is used with specific meanings and definitions in various areas, notably economics, management, international politics etc. Many general definitions of governance in its broader dimension have been given, highlighting the variety of relevant processes and means, such as laws, norms, power or language, as well as the involved actors and levels, for example governments, markets or networks acting over families, tribes, formal or informal organizations or territories (Bevir 2013).

Basically, the concept of governance refers to the processes of interaction and decision-making among the actors involved in a collective problem (Hufty 2011).

The definitions proposed by the International Risk Governance Council (IRGC)[1] are however the most directly relevant for application to risk policy (IRGC 2005):

> **Governance** refers to actions, processes, traditions and institutions by which authority is exercised and decisions are taken and implemented.

> **Risk governance** deals with the identification, assessment, management and communication of risks in a broad context. It includes the totality of actors, rules, conventions, processes and mechanisms and is concerned on how relevant risk information is collected, analyses and communicated, and how management decisions are taken. It allies the principles of good governance that include transparency, effectiveness and efficiency, accountability, strategic focus, sustainability, equity and fairness, respect for the rule of the law and the need for the chosen solution to be politically and legally feasible as well as ethically and publicly acceptable.

It appears, in light of these definitions, that in spite of their common root the terms governance and government have very different meanings. The concept of government refers to the formal institutions and legal powers in a given political

[1] The International Risk Governance Council (IRGC) is an independent non-profit organisation which aims to help improve the understanding and management of risks and opportunities by providing insight into systemic risks that have impacts on human health and safety, on the environment, on the economy and on society at large.

system owning the authority to make formal and binding decisions. Governance refers instead to an inclusive process, which involves a wide range of civil society actors, alongside the institutional ones, in informing and influencing the decision-making process.

7.6.2 Best Practice in Risk Governance

Best practice risk-governance includes some basic steps and activities. First, the risk issue considered is comprehensively explored in a preliminary stage in all its dimensions, with the aim to determine, in addition to the available information on "hard facts", the interests at stake, the perceptions, values, meanings and interests of the various interested parties and of the population exposed. The objective is to complement the analysis of risk, which is science-based and factual, with the understanding of the meaning for the stakeholders of a specific risk, based on values, concerns and the positioning of different groups. This step includes in particular what is sometimes called a "concern assessment". Concern assessment is aimed at systematically exploring the ways in which a risk is appraised by the different groups exposed or interested and the reasons and/or psychological mechanisms at the source of the concerns expressed. At this stage, the stakeholders and interested groups are identified, the channels for dialogue are determined and the appropriate contacts and arrangements are established for a sustained dialogue and collaboration across the preparatory process. The important results of such preliminary step are a broad framing of the risk, the definition of the issues to be submitted to scientific assessment and to be addressed by a comprehensive, systematic and participatory appraisal, and the identification of the relevant data and information to be collected in order to consider all the critical aspects of the risk issue at stake.

Second, an open and participatory approach requires seeking the input of stakeholders on the framing and objectives of the scientific assessment process and the mandate for risk assessors. The objective is to ensure that the issues to be explored are formulated in a way that takes into account all the relevant concerns. Equally, comments and feedback from stakeholders should be actively sought on the draft results of the assessment.

Finally, a two ways dialogue should be ensured on the management options identified and their implementation, the acceptability of residual risks, the risk-risk balance, etc. There is a need to take into account the diversity of interests across stakeholders as well as the different risk-cultures across regions and their development overtime. Plenty of examples show the relevance of such differences, for instance the different attitudes across regions and countries to the presence of GMOs in the food chain.

There is a variety of tools for conducting a pro-active, participatory approach. Written Internet consultations, conferences, workshops and hearings accessible in streaming, focus groups and use of social media may allow reaching a wide audience with limited costs, and collecting extensive input from a broad variety of stakeholders and interested parties.

Along similar lines, the IRGC suggests an approach to risk governance that includes (IRGC 2005):

- Risk pre-assessment, which includes a concern assessment
- Appraisal, which includes the scientific assessment
- Characterisation and evaluation
- Management, including the assessment of options
- Communication

Those steps should not be seen as a sequence. In particular, as shown in Chap. 6, two-ways communication needs to take place across all stages.

The IRGC Risk Governance Framework

IRGC has developed a comprehensive framework for risk governance. *The framework includes five elements*:

- ***Risk Pre-Assessment*** *– early warning and "framing" the risk in order to provide a structured definition of the problem, of how it is framed by different stakeholders, and of how it may best be handled*
- ***Risk Appraisal*** *– combining a scientific risk assessment (of the hazard and its probability) with a systematic concern assessment (of public concerns and perceptions) to provide the knowledge base for subsequent decisions*
- ***Characterisation and Evaluation*** *– in which the scientific data and a thorough understanding of societal values affected by the risk are used to evaluate the risk as acceptable, tolerable (requiring mitigation), or intolerable (unacceptable)*
- ***Risk Management*** *– the actions and remedies needed to avoid, reduce transfer or retain the risk*
- ***Risk Communication*** *– how stakeholders and civil society understand the risk and participate in the risk governance process*

Risk governance therefore incorporates and completes the traditional risk analysis paradigm by including the additional elements of pre-assessment and the wider appraisal, which goes beyond the technical risk assessment and includes also a concern appraisal. While the scientific aspects are well present in the governance scheme, they are integrated into a wider integrative and participatory approach with a wider focus on societal dimensions and aspects of risk. Obviously, not all risks deserve the heavy investment of a structured governance approach. Certain types of risk management measures have a narrow scope and little impact on societal concerns and interests. They may well be dealt with in a more routine and technical perspective. Nevertheless, it is wise to make a careful analysis of the meaning and implications of each risk issue before deciding which approach to follow. The societal impacts of risks are in fact easily underestimated.

Good risk governance is particularly relevant for addressing complexity, uncertainty and ambiguity, three attributes that are often at the source of societal controversies. Let's briefly recall here certain characteristics of risk that we have introduced in Chap. 2, relevant for a participatory approach to risk governance. Risks may be complex because they result from the interaction of many factors in complex technical, economic or other human systems. The underlying science, even if available, may be complex and the amount of data to be processed in order to model the risk-generating processes may be huge. In such cases it may be necessary to reach out to a great number of actors in order to collect the needed information, involve the relevant stakeholders in the identification of effective measures and prepare for their implementation. Since uncertainty is everywhere in risk assessment and decision are therefore taken most often "under uncertainty" rather than "under risk", good risk governance helps ensuring transparency and a wider, consensual view on the state of knowledge and its policy implications and to reach a balance between the relevant risk management principles, notably the precautionary and the proportionality principles. And risks are sometimes "ambiguous" as they are seen differently by, and mean different things to different stakeholders, publics or people. The processes of participatory risk governance should help addressing such different perceptions and perspectives in a more informed, systematic and balanced way.

Risk governance does not stop with the adoption and communication of risk management measures. In the case of complex risks, the concrete application of the measures adopted and the monitoring and evaluation of the relevant developments, and of the results achieved, require the active involvement and collaboration of a variety of actors. The networks, contacts and collaborations built during the preparatory stage may play a very helpful role in that respect. Moreover, measures taken under uncertainty, notably based on the precautionary principle, need to be reviewed and, when new knowledge allows or requires it, reconsidered, reshaped or reinforced.

A participatory approach to risk governance may improve the quality, effectiveness and efficiency of risk policies and regulatory measures. It helps identifying the most relevant issues for the assessment, as well as the stakeholder and public concerns, perceptions and expectations that must be addressed in order to ensure the effectiveness and acceptance of the measures adopted. It helps collecting the data and information that are needed to assess the risk and the management measures in a comprehensive, holistic perspective. It provides information on the interests at stake and helps evaluating the impacts of measures considered.

In a world of rapid technological, economic and organisational change, identifying publicly acceptable and societally sustainable policy choices and effectively managing change is vital. Good governance of risks can make a key contribution to such effective management of change.

Risk governance should however not be seen as a way to "buy in" a broader range of stakeholders or to find ways to paternalistically "convince" the public of the rightness of pre-determined decisions. In order to deliver its benefits, participatory governance must be practiced as a process genuinely contributing to determine the best policies and measures. Good risk governance aims at taking into account the societal context in the assessment/appraisal of risk and in examining the policy and regulatory options.

The added value of participatory risk governance can be summarized as follows:

Benefits of participatory risk governance	
Risk identification, framing and assessment	Earlier and more complete identification of risk issues
	Wider access to and use of diffused relevant information
	Early warning on public sensitivity of certain risk issues
	Identification and assessment of concerns and of interests, views and values at stake
	Framing and focusing risk assessment on the most relevant issues and questions
	More transparent, credible and accepted scientific risk assessment, providing a better view of uncertainties and value-laden assumptions
	More robust common ground among authorities and stakeholders on the factual and scientific facts
	Better chance to achieve a wider consensus on the relevance, relative weight and implications of the various dimensions and attributes of the risk in presence, as well as the benefits and risk-risk balance
Risk management	More complete identification of policy options
	More fairness in risk distribution
	More legitimate assumptions on acceptable or tolerable residual risk, balance between risk and benefits and justification for the introduction of the technology, product, project etc. in question
	Better chance to identify balanced solutions, overcoming/reconciling conflicting interests
	Better, more productive management of confrontational interlocutors trying to interfere with the risk management process
Risk communication	More complete identification of stakeholders and publics to communicate/exchange with, and more direct and effective channels and processes for two-ways communication
	Better and more relevant common ground of knowledge and facts as a basis for dialogue
	More on-substance and less tactical stakeholder dialogue
	Better mutual understanding between stakeholders and with the institutions
	Better chance to dispel misunderstandings, prejudices, biases and correct wrong information and assumptions
	Two-ways, structured communication contributing to better informed, more effective and consensual decisions
Implementation of risk policies and measures	More collaborative and effective approach to policy implementation, notably on complex risk issues
	More timely and more effective feedback on the results, problems, impacts and needs for revision of the measures applied

7.6.3 Potential Risks and Difficulties of Participatory Risk Governance

A participatory governance approach brings benefits to risk policy, but is not without difficulties and risks. The potential problems and drawbacks in that respect overlap with those already discussed in relation to risk communication and dialogue.

Engaging the EU "civil society" in the governance process is far from easy. On the EU level, there is a plethora of associations, NGOs and organised lobbies that present themselves as representatives of their "European" civil society constituencies. Nevertheless, the true representativeness of some of them is not always evident, considering the very different cultures, values, conditions, priorities and interests across the 28 Member States of the EU. Even more difficult is to ensure a balanced representation of all the interests. The main risk of the "participatory" governance is that the process becomes "captive" of two opposite extreme in the range of stakeholder organisations: on the one hand the most powerful and better organised economic and industrial lobbies and, on the other hand, of those parties that are more successful in influencing the public opinion with "scaremongering" and alarmist messages.

Often the debate becomes confused by tactical claims of exaggerated socio-economic impacts of risk prevention measures or, on the opposite side, exaggerations about the nature or level of risk. Moreover, when the doors are open for a variety of interests to have a say in evaluating the risks and pronouncing on the management options, the "official" science has often to compete with pseudo-science, financed or promoted by vested interests, and with the "new data" and "more research" tactics of the "merchants of doubt". Decision makers may then be confused by a cacophony of divergent evaluations being voiced and may in the end tend to ignore the voice of science altogether.

Finally, the concept of a participatory approach to risk governance is sometimes stretched to include scientific risk assessment itself. Certain stakeholders and scholars claim that risk assessment itself should be participatory, performed by experts representing the various interested parties. Such an approach would obviously push risk assessment towards a form of "negotiated science", with the inevitable, serious loss of credibility and weight for science in the decision making process, compared to a truly independent scientific input. An interesting experience was made in that respect in the US in the 1980s in an attempt of ensuring a "balanced" composition of advisory committees based on affiliation instead of independent expertise. The US Occupational Health and Safety Administration- OSHA tried to involve representatives of employers, workers and state health and safety organisations in its advisory committees in the hope of obtaining consensual advice on health and safety risks and related measures. It finally gave up convening committees with such "balanced" composition because they were systematically paralyzed by partisan disagreement between labour and management. Involvement of "civil society" in technical risk assessment is in fact neither desirable nor, in any case, likely to produce any useful result.

In conclusion, participatory governance must not be misinterpreted as a substitute for the political responsibility of the institutions that have got a democratic mandate to take decisions on behalf of citizens in view of the general interest of the public. A participatory approach is necessary in the case of complex, uncertain or ambiguous risks. Its objective is to inform decisions and to establish the best possible conditions for dealing with risks in an effective and efficient manner. Participation of civil society is a difficult exercise. It is seldom perfectly balanced and truly representative of the complex and diverse European society. Once again, the responsibility, the last word on risk management remains with the Institutions that directly or indirectly are accountable to citizens. Industry confederations, NGOs, associations and lobbies are obviously not. Risk governance may well reinforce, but not legitimately replace risk government. Policy makers know well that if things eventually go wrong, they will be the ones held to account.

7.6.4 *What Goes Sometimes Wrong in the Way Governance of Risks Is Organised and Operated?*

Another way to look at the importance of good risk governance is to list things that often go wrong because the process is not well conceived or implemented. We summarise some of such things in the following table, without pretending to be exhaustive and focusing on those aspects of governance that add to the traditional risk analysis approach.

Some frequent defects in risk governance practice	
Pre-assessment	No effective system to anticipate forthcoming risks in certain sectors
	Underestimation of the public sensitivity of certain risk-issues
	Undue delay in recognising relevant risk-issues ("not-in-my-mandate")
	Ignoring societal aspects of risk-issues
	Failing to identify and involve the relevant stakeholders
Appraisal	Appraisal limited to the scientific aspects
	Failure to identify and assess concerns
	Risk assessment launched without a preliminary assessment of all the relevant dimensions of risk, not addressing all the relevant questions
	Risk assessment process not transparently recognising and addressing value-laden aspects and assumptions, uncertainties and other limitations
	Failure to ensure transparency and consultation at the appraisal stage
	Unbalanced representation of interests and stakeholders
Management	Incomplete identification of policy options
	Option identification and assessment not comprehensively considering the diverse perspectives on, and dimensions of the issues at stake
	Biased, unbalanced or incomplete criteria for the assessment of policy options, not properly addressing relevant values and interests
	Application of simplistic models
	Evaluation of option not fairly reflecting the distribution of risks and benefits

(continued)

Some frequent defects in risk governance practice	
Communication	One way communication aimed at "explaining" risk and "selling" to the public unilateral conclusions and decisions
	Lack of clarity on the aims of communication and dialogue processes
	Inadequate means and resources for the management of the communication process
	Lack of mutual trust and respect between the actors involved in the communication
Follow-up	Failure to establish mechanisms to support and monitor implementation of measures, building on stakeholders networks established in the preparatory phase
	Failure to monitor results achieved and problems encountered
	Failure to recognise problems and timely introduce additional or corrective measures, or reconsideration and revision/withdrawal of measures

Summary

1. Managing the variety of risks we are confronted with is in the first instance an individual responsibility. Nevertheless risk regulation is necessary where individuals have no effective means and powers to protect themselves and the goods they value, or no direct, individual interest to prevent them, or they even get a benefit from creating risks which are imposed on others. In particular, protecting the environment and the so-called "commons" requires a collective action that can only be set up and enforced by public institutions.
2. The internal market was and still remains the backbone of the Union. It has been progressively established, and is still being gradually extended and reinforced, through the harmonization of the conditions applicable to the placing on the EU market of goods and the provision of services as well as the free circulation of capitals and workers. Much of the EU harmonization rules are related to health, safety, environmental risks, protecting the interests of consumers and ensuring fair, undistorted competition conditions for producers and service providers. The rationale of most of the EU risk regulation must therefore be seen in the perspective of the functioning of the internal market. This observation has important consequences.
3. Although the focus on systematic impact assessment, transparency and open and inclusive consultation has been enhanced at the beginning of the first decade of this century, it would be wrong to think that older EU policy and regulatory decisions were established in a vacuum of principles and practices ensuring the role of evidence and scientific data or without any ex-ante evaluation and consultation. The EU Institutions were since ever subject to the obligation to motivate all mandatory legal acts adopted. The conditions for the validity of an act have in practice involved the requirement for the EU Institutions, the Commission in the first instance, to collect the relevant information and data and, in the case of risks related to issues with a complex technological or scientific background, to perform an appraisal of the risks in relation to the stated

objectives. Moreover, in preparing its policy on the environment, the European Union is bound to take account of available scientific and technical data.

4. In view of the massive body of harmonization legislation planned for establishing the internal market, a Council Resolution of 1985 launched a "new approach" to technical harmonization and standards that, in order to ensure the free movement of goods, was aimed to harmonise the relevant provisions by establishing only essential safety requirements while leaving to European organisations competent in industrial standardization the task of drawing up non mandatory technical specifications taking into account the latest stage of technology.
5. The European Commission has established a EU Better Regulation approach with the aim to improve the quality of regulation, the transparency of the decision-making process and stakeholder participation across the life cycle of regulation. Better Regulation is organised around three main areas: the ex-ante assessment of new regulation, the consultation of and dialogue with the stakeholders and the public, and the ex-post evaluation, revision and reconsideration of existing legislation. The current policy of the Commission in each area is summarised in a Communication on Better Regulation of 2015.
6. Stakeholder and public consultation is conducted with the interrelated objectives to ensure transparency and accountability of the preparatory process, but also to contribute to the effectiveness of the policies and regulation by collecting relevant evidence. Additionally, the involvement of the interested parties contributes to better understanding, on various levels, of the EU action. The current procedures aim to cover the entire lifecycle of policies and legislation, from the preparatory work and the discussion and adoption of a decision, to the ex-post evaluation of its application. The Impact Assessment approach is based on integrated consideration of economic social and environmental impacts. The assessment of risk-related regulation must take into account results of the risk assessment made by EU specialized bodies. Reference is made in the Better Regulation guidelines to the commitment to use the best available evidence and science. A new stress is put on the need to monitor and assess existing plans and measures and to factor the results in the assessment of new initiatives. Finally, the REFIT Programme is aimed at supporting the continuous adaptation of EU legislation, in light of its results, effectiveness and efficiency, any problems encountered or created in its implementation, new needs and the changing economic, social, natural and environmental context. Action may be needed in order to adapt, simplify or simply withdraw certain policies or regulatory instruments.
7. Better Regulation has been welcome in principle by the EU Institutions and the Member States, by part of the stakeholders, notably business, by trade partners and international organisations, notably the OECD and by part of the observers and scholars. It has however been regarded with suspicion by most of the NGOs and other organisations focusing on health, safety, environmental, consumer and worker protection, which have seen it has a business-lobby driven approach aimed at deregulation, the lowering of European protection standards, com-

pression of democratic control and excessive influence of vested interests over public policies. On the EU level, the European Commission has consistently insisted on a balanced approach, rejecting allegations that Better Regulation would allow short term and unilateral business interests to prevail over other public interest imperatives, political choice and democratic accountability. The Better Regulation approach poses some substantive problems that can only be solved by clarifying its objectives and its role in the decision-making process, moderating expectations, ensuring a really balanced, integrated and credible approach, balancing the resources invested and the added value of the process in a pragmatic and realistic fashion.

8. The best use and the real added value of "science" in the regulatory process is to help structuring the multi-criteria assessment and appraisal of risks, and to inform the process of discussion, deliberation and decision. Science becomes suspect for part of the stakeholders and policy makers when it is presented as a substitute for a broader "political" mediation between the facts, uncertainties, interests and values that most often intervene in dealing with controversial risk issues. Science is in fact sometimes presented as a means "to rescue policy from short-sighted influence peddling and power politics".
9. Factual and scientific evidence is key for supporting rational decision making. However, a simplistic insistence on the concept of "evidence-based" decision making often fails to recognise the complexity and multidimensional nature of the political decisions. Politicians, who in the end have to assume the responsibility of decisions, have their own perspective of what is "effective". They need to take into account divergent interests and the weight of public expectations, perceptions and values.
10. Rational and efficient regulation should try to anticipate risks, notably based on the Precautionary Principle. But an excess of precaution poses the risk to over-regulate, putting a brake on innovation in the absence of sufficient data and adequate scientific knowledge of the potential adverse effects at the early stages of development and use of a new technology. While in some cases precaution may justify a stringent preventive approach (notably: no data, no market), the needs of innovation can often be better addressed through an adaptive approach. The first steps of an adaptive strategy may focus on knowledge improvement and early action focused on preventing possible irreversible damage. Adaptive regulation involves continuous monitoring of its effects, collection of data and parallel research on plausible potential risks. The initial constraints on the development and use of the technology or products considered should be proportional and limited to the objective to mitigate the potential risks and to prevent serious, irreversible damage, should the concerns materialise. Measures should be timely reinforced (or waived), as appropriate, as soon as new evidence becomes available. Adaptive regulation requires a cultural change both on the side of regulators and by industry and promoters. It can work well if all stakeholders collaborate constructively.
11. The difficulty of innovation control was well expressed by David Collingridge of the Technology Policy Unit at the University of Aston, already in 1980, in his

book *The Social Control of Technology*, with the so-called *Collingridge dilemma* stating that it is impossible to foresee the need for change at the early stage of a development, when change is still easy, while when the need for change becomes apparent, changing course is expensive, difficult and time-consuming. There is a need to incorporate the risk analysis paradigm into innovation development at an early stage, with the appropriate instruments and resources. Development of innovation and production of data and scientific knowledge on its possible adverse effects should proceed in parallel.

12. Because of the nature of risks, which most often involve both "technical" and "societal" aspects, openness and participation by the civil society are particularly important in order to ensure effective management decisions. Therefore, the risk analysis paradigm needs to be framed in its wider and complex societal context. To that aim, it is necessary to develop and organise a wider participatory framework integrating, in an ordered, balanced and transparent way, the civil society's contribution with the EU institutional decision making process. The concept of "government" must be broadened towards that of "governance". The risk issue considered is explored in a preliminary stage in all its dimensions, with the aim to determine, in addition to the available information on "hard facts", the interests at stake, the perceptions, concerns, values and interests of the various interested parties and of the population exposed. This step includes in particular what is sometimes called a "concern assessment" aimed at systematically exploring the ways in which a risk is appraised by the different groups exposed and the reasons and/or psychological mechanisms at the source of the concerns expressed.
13. An open and participatory approach requires seeking the input of stakeholders on the framing and objectives of the scientific assessment process and on the mandate for risk assessors. The objective is to ensure that the issues to be explored are formulated in a way that takes into account all the relevant concerns. Equally, comments and feedback from stakeholders should be actively sought on the draft results of the assessment. Finally, a two ways dialogue should be ensured on the management options identified and their implementation, the acceptability of residual risks, the risk-risk balance, etc.

References

Bevir, M. (2013). *Governance: A very short introduction*. Oxford University Press, Oxford, UK, 2013.

Collingridge, D. (1980). *The social control of technology*. New York: St. Martin's Press; London: Frances Pinter, 1980.

European Commission. (2001). *White paper on European Governance*. COM 428, 2001.

European Commission. (2015). *Better regulation for better results*. COM 215, 2015-1.

European Commission. (2015). *Proposal for an institutional agreement on better regulation*. COM 216, 2015-2.

European Commission. (2015). *Regulatory scrutiny board mission, tasks and staff.* C8 3262, 2015-3.
European Commission. (2015). *Decision establishing the REFIT platform.* C b3261, 2015-4.
European Commission. (2015). *The REFIT platform structure and functioning*, C. 3260, 2015-5.
European Council. (1985). *Resolution 85/C136/01.*
European Council. (2000). *Presidency conclusions 23–24 March 2000.*
European Court of Justice. (1984). *Case 258/84.*
European Court of Justice. (1987). *Joined cases. 279, 280, 285, 286/84.*
European Court of Justice. (2010). *Case C-343/09.*
European Court of Justice (General Court). (2015). *CaseT-521/14.*
European Union. (2002). *Regulation (EC) no 178/2000 of the European Parliament and the Council laying down the general principles and requirements of food law, establishing the Food Safety Authority and laying down procedures in matters of food safety.*
European Union. (2006). *Regulation (EC) no 1907/2006 of the European Parliament and the Council concerning the registration, evaluation, authorisation and restriction of chemicals.*
Hufty, M. (2011). Investigating policy processes: The Governance Analytical Framework (GAF). In U. Wiesmann, H. Hurni, et al. (Eds.), *Research for sustainable development: Foundations, experiences, and perspectives* (pp. 403–424). Bern: Geographica Bernensia.
IRGC (International Risk Governance Council). (2005). *Towards an integrative approach.*
Majone, G. (1996). *Regulating Europe.* London: Routledge.
Mandelkern Group on Better Regulation. (2001). *Final report.*
Redaelli, C. M. (2005). *What does regulatory impact assessment mean in Europe.* AEI-Brookings Joint Center for Regulatory Studies, Washington, D.C.
The Evaluation Partnership. (2007). *Evaluation of the commission's impact assessment system.*
US NRC (National Research Council). (2012). *Using science as evidence in public policy.* The National Academies Press, Washington, D.C., 2012.

Chapter 8
Ensuring and Enhancing the Quality and Utility of Risk Assessment

Abstract There are several procedural and substantive conditions for ensuring the utility of risk assessment in the decision-making process. The specific nature, role and limits of risk assessment should be understood and respected by all the actors. Policy makers should not place unrealistic expectations in the power of science to deliver clear-cut "truths" on risk. They should not "hide" behind science when taking decisions that are, at least partly, motivated by public requests, economic constraints or societal preferences. There is a need for an internationally recognised, wider framework of principles and criteria for regulatory decisions, incorporating the entire range of "imperative requirements" that are relevant for the comprehensive appraisal of risks. Scientists should be careful to remain within the limits of their scientific task and managers and policy makers must not interfere with the integrity of risk assessment. Nevertheless, the separation between assessment and management of risks should not prevent a close and sustained dialogue between managers and assessors. There is a need for a "risk assessment policy", jointly established by managers and assessors, endorsed by the political decision makers, integrating the principles, criteria and good practices that have been established by the EU legislature and by the risk assessment bodies. The quality and utility of risk assessment depend on several factors: the competence and independence of the scientists delivering the advice, the availability of adequate data, the quality and transparency of the procedures, processes and methods applied, but also the clarity of the expression of risks and of the associated uncertainties. This Chapter discusses the issues, problems, best practices and areas for improvement for ensuring and enhancing the quality and utility of risk assessment.

8.1 Introduction

We have introduced in Chap. 4 risk assessment in its technical dimension, providing a detailed description of the approach and methods used to assess certain categories of risks, notably the chemical risks. In spite of the many considerations that we have developed in previous Chapters on the multidimensional nature of the decision-making process on risks, it should be clear that risk assessment remains at the core of the EU approach to risk analysis, policy, regulation and governance. As a science-based decision-aid tool it involves the use of complex technical

B. Delogu, *Risk Analysis and Governance in EU Policy Making and Regulation*,
DOI 10.1007/978-3-319-30822-7_8

methods, adapted to the specific risk-issue considered. Assessment methods have been developed by scientists and practitioners and are continuously improved and adapted to scientific and technological change. We have seen that under the EU approach, and in line with international best practice, risk assessment is functionally separated from risk management and policy making. On the positive side, such functional separation contributes to preserving the independence and integrity of the scientific dimension of the risk analysis paradigm. However, we have already noted in previous Chapters that a too strict separation between the scientific and the policy making processes may have some important drawbacks. In particular, due to the political context in which the risk analysis paradigm was introduced in the wake of the post-BSE reform and to the need to respond to concerns on the independence of assessors, the EU Institutions did not define a risk assessment policy, considering that policy makers should not intervene in the setting of risk assessment procedures and methods.

Therefore, although the EU attributes to independent scientific risk assessment a key role in the EU policy making process, there has not been so far a structured dialogue between EU policy makers and risk assessment bodies, which is needed for defining a mutually agreed risk assessment policy. As a consequence, there is no formal document endorsed by the EU Institutions setting a horizontal framework of principles, procedures and criteria that risk assessors are requested to apply. Risk assessors have developed their own models and approaches, with policy-relevant implications, which however are not shared with, and endorsed by policy makers. Criticisms are sometimes expressed by policy makers and stakeholders on the assumptions and value-laden criteria applied, and choices made by risk assessors. We will see more in detail in this Chapter how certain methodological and procedural weaknesses may affect the way in which risk assessment is structured and how its results are expressed. Such weaknesses limit the utility of risk assessment in informing decision making. This Chapter discusses the most important aspects on which risk assessment may still be improved in view of optimizing its transparency, quality and utility for risk management and policy making.

8.2 Risk Assessment's Role in Support of Management, Policy and Governance

The role of risk assessment in a well-developed risk analysis system is to provide risk managers, decision makers, stakeholders and the public with information on the characteristics, severity, likelihood and distribution of the risks related to products, technologies, processes, natural events, practices, human behaviours etc., based on the best available scientific knowledge, the best available data and proper and recognised assessment methods and procedures. We have discussed in detail the importance of a systematic approach to uncertainties, the need for transparent and effective expression and communication of any such uncertainty and of transparency and clarity on the meaning and possible limitations of the assessment results.

The scientific quality of the assessment is of course of the utmost importance. Nevertheless, the utility and practical weight of risk assessment in the decision making and risk governance processes also depend on several additional factors.

Clarity on the Nature and Limits of Risk Assessment First, the specific role of risk assessment as part of the decision-making preparatory process should be clearly understood and its integrity respected by all the actors and users involved, including the assessors themselves. The starting point for improving the utility of risk assessment is in fact to ensure a better understanding by all the actors and stakeholders, across all the risk analysis stages, of both the potential and the limits of science in determining the reality, probability, severity, extent and distribution of risks.

Well-designed and well-conducted risk assessment provides structured information on risks, based on best available data and evidence, applying conventional models, assumptions and defaults, which most often include expert judgment and extrapolations. Risk assessment supports ordered and rational reasoning about the existence, nature, size and potential impact of risks, the probability of damage occurring and the uncertainties involved. However, we have repeatedly shown that risk assessment is often affected by significant limitations. The challenge for risk assessors is to provide risk managers with helpful information in spite of all the practical and methodological difficulties. To that aim, it is vital that decision makers are informed on the basics of risk assessment methods and are well aware of what risk assessment can and cannot deliver.

Decision makers often expect a "scientific" guarantee of safety or a deterministic quantification of risks that science is rarely in a position to deliver in case of complex or controversial issues. Opinions expressly characterised by uncertainty or including minority conclusions cause deception and scepticism among decision makers who may be surprised to discover that "science" is not always in a position to deliver a consensual "truth" about the risks considered. The full added value of science-based risk assessment cannot materialise in the decision making process if risk managers and decision makers are not made aware of the nature and limits of the assessment tools and are not in a position to properly interpret and use the results of an assessment. To that aim, risk assessment reports should transparently present all important assumptions underlying the conclusions reached as well as the limitations of such conclusions. The reasons for divergent opinions should be objectively analysed and explained. A sustained dialogue and a close interaction between assessors and managers across all the stages of a risk assessment can help ensuring a better understanding by all the actors of the risk assessment process and of its conclusions. Organising such a dialogue under the EU conditions requires a systematic approach, involving all the separate institutional layers of the European decision making process.

The Role of Science in Informing Decisions Once the nature and limits of risk assessment are transparently explained and well understood, it is important to be clear about the intrinsic nature of the role of science in informing decisions. We have shown in previous Chapters that, although scientific risk assessment provides a key input into the broader assessment of risk management options, a comprehensive appraisal needs to be multi-dimensional and multi-criteria: costs and benefits, legal aspects, industrial, social and administrative aspects as well as ethical considerations must necessarily play a part, alongside the results from risk assessment, within a comprehensive and balanced approach.

Differently from jurisdictions where regulatory powers are attributed to specialised, technical agencies, in the EU political institutions decide upon risk-management measures. The political decision makers are democratically accountable to voters and need to take into account the wide variety of diverse interests that are at stake. We have discussed in Chap. 7 how the concept of "science-based policy making" may be misleading if it is narrowly interpreted as implying a de-facto delegation of decisions to scientific bodies or suggesting that "science" alone, and in particular science-based risk assessment, may determine the "best" management option. For example, allowing or prohibiting the use of a new technology or the marketing of a product, or defining conditions for such use or marketing, or authorizing a new project, obviously involve considerations which are not exclusively related to risks, but also need to be factored in the assessment alongside the risk assessment findings.

It must be noted that international trade agreements have introduced constraints on the public policy objectives that may be invoked to justify measures that create trade barriers. Unfortunately, there is not yet a comprehensive framework of internationally recognised criteria for guiding decision making, integrating the plurality of legitimate objectives, interests and criteria that are relevant in practice for political decisions. Sometimes, health and environmental risks are invoked to justify measures pursuing other public policy objectives or intended to respond to democratically expressed public concerns or preferences. Policy makers sometimes tend to "hide" behind science, while taking decisions that are motivated by considerations that have little to do with scientific evidence on health, safety and environmental risks. Disagreement on health and the environmental risks is sometimes at the core of trade disputes, although in fact other reasons, not related to scientific risk assessment, have played a major role in determining the disputed measures. Stakeholders frequently "interpret" scientific advice as it is convenient for advancing their agendas. And sometimes "official" science is criticized with weak pseudo-scientific arguments, when it presents inconvenient conclusions which are seen to go against a course of action promoted on grounds of interest, values, ideology, societal preferences etc. In all these cases, science is stretched to cover objectives and interests that are not transparently made explicit.

When health and environmental risks are invoked in order to justify measures that are in fact aimed at pursuing different policy objectives or interests, the integrity and credibility of the risk analysis paradigm suffers. There is a need for a more comprehensive framework of recognised, transparent criteria for the assessment of policy options.

Scientists themselves sometimes contribute to the misconception of science as the only acceptable source of legitimacy that decision makers, and all those interested in governance of a given risk, may invoke. Scientists should be careful to show that they accept that science alone cannot decide the "right" course of action, irrespective of economic and political aspects, interests and values of the parties involved and the other relevant societal aspects. Patronising attitudes, critical statements on public or stakeholders concerns, often dismissed by experts as "irrational", impatience with "politicians" who are seen with some contempt as just acting on "political agendas", are all attitudes, which unfortunately we witness from time to time, that are incompatible with a good-practice code of conduct for scientists. Assessors should carefully remain within the boundaries of their task, which is limited to providing scientific expertise. They should refrain from entering the policy debate, where values, costs, benefits, interests play a legitimate and important part. In order to preserve the integrity and specific utility of their role, scientists should aim to support the informed thinking of *all* the parties involved, avoiding confrontational positions and showing full respect for all different views in order to be recognised by all stakeholders as neutral and independent. They should help making a clear distinction between the scientific aspects and other factors that stakeholders may consider to be relevant and legitimate. The best way to preserve such distinction and to protect the role of science is not to pretend that science should necessarily have the last word in a policy debate. Otherwise, as it too often happens, part of the actors in the risk governance process will altogether dismiss science as being unreliable, partisan, biased, uncertain and irrelevant etc.

An excellent example of good practice in that respect was provided by Peter Gluckman, New Zealand's chief science adviser, who before being appointed in 2009 as chief adviser had pronounced in favour of the addition of folate to flour-based products as a means to prevent neural-tube defects in new-borns, but when the government decided not to require such addition, as appointed chief adviser declared in an interview that a democratic government could not easily ignore the overwhelming public concern about the food supply in spite of what he considered "strong scientific evidence". As he put in an article (Gluckman 2014):

> "I have come to understand that the primary functions and greatest challenges for a science adviser are providing advice not on straightforward scientific matters, but instead on issues that have the hallmarks of what have been called post-normal science. These issues are urgent and of high public and political concern; the people involved hold strong positions based on their values, and the science is complex, incomplete and uncertain. Diverse meanings and understandings of risks and trade-offs dominate..................
>
> ....Science advice is about presenting a rigorous analysis of what we do and do not know. Alone, it does not make policy. There are many other appropriate inputs to policy, including fiscal considerations and public opinion. Policy-makers and elected officials rightly guard their responsibility to define policy – and this means choosing between options with different trade-offs. This is not the domain of a science adviser. Being explicit about this has eased my capacity to establish and sustain trust broadly across government and the policy community."

8.3 A Better and Closer Link Between Assessment and Management of Risks

The way in which EU risk managers formulate their risk assessment questions depends on the applicable regulatory framework, the nature of the issue and the specific policy objectives for the relevant case. Sometimes, questions are very narrowly formulated (for instance, in the case of questions related to specific ingredients in cosmetics) while other requests are related to very broad risk issues (like for example the potential risks of nanomaterials or endocrine disrupting substances).

When opinions are requested in order to support decisions on implementing or delegated measures, which are taken within the framework of an existing regulation, the mandates to scientists tend to repeat literally the wording of the legislation. In such cases, although the established doctrine would require that risk assessors should not pronounce on management measures, the questions are sometimes formulated in such a way that the answers provided by the assessors, together with the provisions of the applicable legislation, leave very little room for interpretation and have therefore very transparent and direct management implications. For example, if the applicable legislation states that products or substances belonging to a certain category may only be placed on the market if they are "safe", asking scientist to say whether a particular product or substance is safe puts the assessors in the position of actually determining with their answer the resulting regulatory decision. Risk managers try to maximize the legal security of the measures that they take by adhering in their mandates to risk assessors to the exact wording of the relevant regulatory provisions. However, this type of direct link between assessment and management is inappropriate on two grounds: often it pushes scientists to oversimplify the expression of risk (for instance to reply with a "safe, not-safe" or similar statement); moreover, it quite directly involves scientists in risk management.

Curiously, while such confusion of roles is often accepted, it is not common practice to explore possible options for action before launching a risk assessment and ask assessors to evaluate the risks under the corresponding scenarios. Associating risk assessors to the evaluation of policy options still raises suspicion of involvement of scientists in risk management. Nevertheless, comparing the risk reduction potential of various management options would facilitate the identification of the most efficient course of action. Currently, assessors are usually requested to assess only risks associated with a baseline scenario. Given the complexity of the EU decision making process, iterative consultations of risk assessment bodies are difficult to make in practice. As a consequence, reliable information on the risk reduction potential of the different options that are considered is often missing. That makes it difficult to evaluate the efficiency and the possible best modulation of the options considered. In that respect, a more effective approach is recommended by the US National Research Council (NRC) in its report "Science and Decisions", which has updated and completed in 2009 the already mentioned seminal NRC "Red Book" of 1982, one of the historical references for the EU risk analysis system. The NRC proposes a more comprehensive and articulated framework for risk analysis (US NRC 2009) under which risk assessors should collect and analyse, with appropriate methods, data and knowledge about threats to human health and to the environment only after risk managers have performed an evaluation of the options available to manage the problems at hand and defined the questions that risk assessment should address against the background of the risk management issues so identified. "Science and Decisions" states that the idea that assessors should be shielded from the specific decision-making issues that their analyses are intended to support derives from misinterpretations of the Red Book. In order to be useful and ensure its added-value, risk assessment should be oriented to help discriminate among risk-management options. Instead of focusing on isolated aspects, such as, for example, the risk profile of a chemical substance, risk assessment information should be presented in such a way as to help comparing the worth of competing management options.

Comparison between different policy or regulatory options may also be necessary in order to evaluate their respective risk-risk balance. As we have extensively discussed in Chap. 5, measures restricting or banning the introduction of a given technology, product or substance on grounds of risks may create ancillary risks that should be compared with those that are prevented or mitigated by the measures in question. It may be important to determine the regulatory approach optimizing the balance between the residual and the ancillary risks. For that aim, managers should support a comprehensive, comparative assessment of the risk balance for all the options considered.

More in general, risk assessment and the broader assessment of policy options, that in the EU decision making process take place as Impact Assessment, should be part of an integrated assessment approach. For policies and measures that involve consideration of risks, risks should be evaluated together with costs and benefits, in parallel, for the various possible regulatory options. Again, it is necessary to that aim to avoid a too rigid separation between risk assessment and management: a systematic dialogue and a close collaboration between the assessment bodies and managers should take place across the various stages of policy analysis, while preserving the integrity of risk assessment.

8.4 Ensuring a More Consistent, Transparent and Policy-Relevant Expression of Risks

In the absence of an EU policy for risk assessment endorsed by the Institutions, there is no consistent guidance so far across the many EU policy sectors on the way in which risks should be expressed. In practice, expression of risks is very diverse going from the very simplistic ("safe/not-safe") to the very technical (Margin of Safety referred to a NOAEL or a Benchmark Dose), and from the generic/qualitative characterisation to the use of quantitative metrics. Sometimes, risks are expressed in ways that policy makers may find it difficult to understand or interpret.

Policies and specific measures addressing risks are intended to protect goods that society values, in particular health, the environment, biodiversity or property. Nevertheless, when risks are expressed in technical terms, it is difficult for decision makers to obtain from such expressions a clear picture of what is at stake and to see what benefits, in terms of protection of social goods, could be achieved by the options considered. For example, an assessment report that informs decision makers that, on the basis of animal data for a certain end point and as a result of a single point exposure assessment, a worst-case "margin of safety" of 80 was estimated, does not really help them in appraising the actual nature and size of the public health problem at stake and the health benefits of possible policy options. On the other hand, qualitative phrases stating that the exposure to a risk factor "gives reasons for concern" or, on the contrary, "poses negligible risks", widely used in risk assessment of chemicals, carry a very poor and potentially misleading message. Much more helpful would be to provide, whenever possible, estimates of how many additional cases of a certain pathology could be expected in a given population; how such disease incidence would be distributed across the population; which vulnerable groups would be most affected etc., together with a narrative helping managers to properly interpret the quantitative information provided. There are various reasons why risks are often expressed in terms so distant from what really matters. As we have seen, one is the wish of risk managers and decision makers to protect the legal security of their decisions by adopting a simple and narrow formulation of their questions closer to the terms of the applicable legislation. Risk assessors in turn may not be prepared to deviate from established practice and to replace the familiar, conventional and technical terminology with a narrative based on the evaluation of potential impacts of risks and of the benefits of management options. And, finally, data and methods for expressing risks in terms of relevant public goods may not be available or may require substantial resources or time to obtain.

In an opinion adopted in March 2013 on "Making Risk Assessment More Relevant for Risk Management", the Commission Scientific Committees SCENHIR, SCHER and SCCS have observed (SCCS et al. 2013):

> Current approaches to the assessment of health and environmental risks frequently result in a variety of technical expressions of risks, based on consideration of endpoints, biological responses or other technical parameters that are sometimes only very indirectly related to the protection objectives pursued by risk managers and policy makers. On the other hand, in posing the questions, risk managers do not always provide an appropriate framework, notably by specifying the protection policy objectives in a manner that would allow the risk assessors to deliver outputs that are readily usable and leave no room for misinterpretations. As a result, interpretation of expressions used in risk assessment reports may be problematic for risk managers and the public, open to misunderstanding and distortion, and difficult to communicate. Moreover, risk assessment reports rarely address in a direct, systematic and transparent way the risk-risk and, where it is possible to express risk and benefits by an appropriate metrics, risk-benefit balance issues which may arise in the

(continued)

specific cases examined. Finally, risk assessment methods, procedures and expression of results are rarely coordinated with the cost-benefit or, more generally, multi-criteria assessment that risk managers and policy makers also need to inform their decisions. A crucial challenge in this process is the possibility (still to be determined) to establish and standardise approaches and methodologies to measure and quantify risks, benefits and costs so that weighing of these parameters in the risk assessment and risk management processes can be done in a comparable and meaningful manner. As a result of the lack of such approaches, comparisons of non-standardised parameters (e.g. potential health benefit versus socio-economic costs) vary considerably depending on the underlying assumptions used thereby limiting the value of these comparisons in policy making. Risk assessors often fill the gap between the technical parameters resulting from the process (e.g.: a margin of safety) and the risk managers' questions by "interpreting" themselves the risk assessment through more or less conventional expressions for "grading" the risk. When that happens outside a pre-defined and agreed scheme to frame risk expression, misunderstandings and confusion of roles are possible.

8.5 Expressing Uncertainties in a Transparent, Meaningful and Understandable Way

We have seen that risk assessment is most often affected by various types of uncertainties. Uncertainties may be related to the data available, the models used, the parameters introduced into the models, the assumed defaults etc. In addition, the estimated level of risk may be modulated across the population considered as a function of the variability in the exposure, vulnerability and other relevant aspects. Transparency about uncertainty and variability is an essential requirement of best-practice risk assessment. Various approaches and methods have been defined and are applied by the EU risk assessment bodies to manage uncertainty. Nevertheless no consistent approach is available for now across sectors for the expression of uncertainties. A sustained dialogue is necessary between EU assessors and managers on how to present uncertainties, which expressions and metrics would best meet the needs of managers and how to minimise the risk of misunderstandings on the interpretation of uncertainties.

Most often, assessors use qualitative expressions for characterising uncertainties in risk assessment reports (very likely, likely, not very likely etc.). Unfortunately, in the absence of definitions, such expressions may be interpreted in very different ways. An interesting approach has been developed in that respect by the International Panel on Climate Change, associating a qualitative grading of uncertainties with the indication of the corresponding quantitative ranges (IPCC 2010) (Table 8.1):

In the IPCC approach, the grading of uncertainty was accompanied by a parallel grading of the confidence in the findings presented (Table 8.2), replaced by a qualitative approach in the fifth report:

We have discussed in Chap. 4 the approach to uncertainty in risk assessment and to its expression, introducing in particular the tiered approach. The IPCC grading is an interesting attempt to define more rigorously the meaning of qualitative

Table 8.1 IPCC likelihood scale

Terminology	Likelihood of the occurrence/outcome
Virtually certain	>99 % probability of occurrence
Very likely	>90 % probability
Likely	>66 % probability
About as likely as not	33–66 % probability
Unlikely	<33 % probability
Very unlikely	<10 % probability
Exceptionally unlikely	<1 % probability

Table 8.2 IPCC quantitatively calibrated levels of confidence

Terminology	Degree of confidence in being correct
Very high confidence	At least 9 out of 10 chance of being correct
High confidence	About 8 out of 10 chance
Medium confidence	About 5 out of 10 chance
Low confidence	About 2 out of 10 chance
Very low confidence	Less than 1 out of 10 chance

expressions of likelihood and confidence in order to promote consistency in their use and a more uniform understanding of such expressions by users of the IPCC assessment reports.

8.6 The EU Core Principles of Risk Assessment

Following the experience of risk-crises, notably the BSE crisis, at the end of the 1990s the EU has set excellence, independence and transparency as core principles of its reorganized risk assessment system. Additionally, the Commission "Better Regulation" initiative has enhanced the focus on responsiveness and openness in all policy and regulatory areas. These principles have been incorporated into the rules and practical functioning of the EU risk assessment bodies.

Excellence refers to the overall quality of the scientific advice. It depends in the first instance on the competence of the scientists providing such advice, but also on the methods, processes, data used, as well as the quality of the working procedures and of the communication.

Independence includes both the independence of science from policy considerations and political interests, as well as the functional separation between risk assessment and risk management responsibilities, and the independence from any vested interest of lobbies and the distance from political and advocacy positions.

Transparency involves both the transparency of the procedures and working methods applied and the full, pro-active disclosure and communication of the scientific processes, data, assumptions, methods and criteria used for reaching the risk assessment conclusions. Transparency also requires disclosure of minority opinions.

Finally, the objectives of openness and responsiveness have been put in practice through procedures for pro-active public consultation and stakeholder dialogue, with a view to get all useful information helping to shape the questions to risk assessors and to address the scientific issues in the most appropriate way in order to respond to stakeholders' questions and concerns.

We have discussed in detail the objectives and good practices on openness and responsiveness in the Chapters on risk communication and risk governance. We discuss hereafter in more detail the implementation of the core principles of excellence, independence and transparency.

8.6.1 Excellence

In its communication of 1997 on consumer health and food safety (European Commission 1997), explaining the new organisation of risk analysis and control for ensuring health and safety of consumers, the Commission stated that scientific qualification and competence are the most important criteria for selecting members and experts of the Scientific Committees. The Commission aims at ensuring the scientific excellence of its entire advisory-support structure. In a communication of 2002 on the collection and use of expertise by the Commission (European Commission 2002), reference was made again to the importance of excellence of the scientists involved.

Ensuring scientific "excellence" is in fact a complex exercise that does not stop with the selection of members of a scientific committee. The issues on which high quality scientific advice is most needed are often complex, multidisciplinary, new or emerging, and are frequently characterised by uncertainty and scarcity of pertinent data for risk assessment. "Competence" can only be ensured by providing for adequate collective expertise, including through the ad hoc involvement of experts external to the relevant standing committees or panels. A wide range of disciplines are represented at high scientific level in the EU permanent committees and in the agencies providing risk assessment advice. However, the in-house expertise of the committees and of risk assessment bodies may not always be sufficient to deal with the wide range of issues encountered in risk assessment since the best knowledge needed for the assessment of new or emerging risks may often only be found in specialised research institutions or in industry. Accessing and involving such highly specialised expertise may be difficult as researchers working on cutting-edge issues have little time and motivation for risk assessment and often are not familiar with the risk assessment techniques. Moreover, the involvement of industry experts may not be compatible with the requirement of independence, and may be constrained by imperatives of commercial confidentiality.

Very qualified and knowledgeable scientists are needed as members of risk assessment committees. However, it is easier for end-of-career or retired academics to devote sufficient time to the demanding activities of a scientific advice body; young scientists, who are often more up-to-date on the "hot" issues of the moment, are not necessarily interested in this type of activity and often could not afford to participate because of time constraints. That difficulty is aggravated by the practice of not providing adequate compensation to the scientific or research institutions that accept to put one of their scientists part-time at the disposal of an independent advice body. Finally, the few scientists with the appropriate profile on topical issues are often disputed between several national and international advice bodies and could not respond to all demands.

In order to deal with these difficulties, a great deal of flexibility is necessary in the organisation of scientific expertise. Current structures may more and more need to adapt their working procedures in order to deliver, at the required level of excellence, advice on increasingly complex scientific issues, and to respond to a heavier load of requests. Increased networking and structured collaboration between national and international bodies, within the EU and with non-EU countries, might also help in that respect.

Progress may also be achieved by broadening the range of expert profiles represented in risk assessment activities. Traditional academic criteria prevail in the selection of committees' members (peer reviewed publications, academic titles obtained and positions held etc.). However, other types of non-academic qualifications may be needed in fields like for example engineering, measurement methods, medicine. Certain profiles are often considered irrelevant because they relate to practical expertise that can contribute mainly anecdotal (but not necessarily irrelevant) knowledge. A more comprehensive representation of different types of expertise may well contribute to the quality of the advice in certain domains.

Excellence of the advice provided depends on a variety of additional factors, on top of the scientific excellence and appropriate expertise of the scientists and experts involved. Independence and transparency are discussed separately. Here we refer to the availability and quality of data.

Under the EU organisation of risk assessment, EU scientific bodies provide risk assessment advice either on the basis of data-files provided by industry under regulatory requirements or, following requests from the EU Institutions, in light of knowledge of scientists and experts and ad hoc identification and collection of relevant data. A structured technical support for researching data and a closer link with the EU and national research resources would contribute to increase the quality and robustness of the advice produced. In case of opinions based on industry files, a better access to all relevant data from other sources might often allow a more comprehensive and informed critical scrutiny. Public calls represent an easy tool for collecting data, but often require a significant effort for filtering and selecting the relevant ones.

There is no single model across the EU advice bodies to ensure excellence. Certain risk assessment bodies, notably the European Commission Scientific Committees and EFSA, launch open calls for application to membership as committee or panel member, followed by structured assessment and selection procedures. Other bodies, like in particular ECHA for its Risk Assessment Committee, rely upon nominations by the Member States. In general, there is a tension between the principles of excellence and independence.

8.6.2 Independence

Independence of scientific advice is a basic principle of the EU risk analysis approach. Assessors must be in a position to perform their tasks without interference or influence from policy makers or any sort of vested interest. That principle requires in the first instance the functional separation between risk assessment and risk management. As we have discussed previously, such separation must not prevent the necessary dialogue between managers and assessors. We have already highlighted the importance of better integrating assessment and management of risks: although it is important that scientists are protected from political pressures, the practical value and contribution of science is greatly reduced in the absence of a sustained dialogue between assessors and managers.

Independence from vested interests also requires that all those playing a role in determining the assessment conclusions be free from the influence of economic or other personal interests. Independence from industry is particularly relevant when considering risks related to products, processes or technologies. EU bodies have developed systematic approaches to ensure that opinions be free from any undue influence of vested interests. A very good and well documented example of such systematic approach is provided by the rules and procedures established by EFSA. Members and experts participating in risk assessment activities are required to declare any relevant interests. The interests are then evaluated, normally by the staff ensuring the organisation of the work together with the Chair of the committee or panel where the work takes place. The evaluation is made in relation to the tasks of the group and the role of the member or expert in question. Based on the results of the assessment, it is determined whether there is an interest potentially conflicting with the required independence. It is then decided whether, to what extent and how the member or expert may participate in all or some of the activities of the group.

Rules on independence are constantly reviewed and have been gradually improved, completed and reinforced. There are some slight differences between the ways in which decisions on the consequences of a conflict of interest are taken in the various EU bodies. EFSA has established very detailed guidance while other bodies relies more on a case-by-case assessment by the Chair, the secretariat and the relevant committee, within the framework of more or less detailed guidelines. In general, too rigid criteria on independence would make it impossible to access the best expertise in certain areas of science and technology, where collaboration between research bodies, academia and industry is so vital and developed that it is practically impossible to find knowledgeable experts on certain subjects who have not collaborated with industry or participated in projects at least partly financed by industry. Moreover, there is a real risk that the enquires into possible links with economic or industrial interests, which are becoming more and more intrusive, sometimes also involving the relatives of the scientists participating in risk assessment work, may demotivate scientists who, in the EU system, provide their expertise largely pro-bono.

Finally, certain organisations, commentators and interested parties that disagree with the conclusions of a risk assessment body, frequently criticise unfairly individual scientists on a personal basis, often expressing groundless and even offensive doubts on their independence, integrity and neutrality. Such unfair practices may also contribute to discourage from contributing to risk assessment work those scientists who are most involved in cutting-edge research, involving collaboration with, or financing by industry.

In ensuring the objective of independence, a distinction must be done between the presence of an interest and a conflict of interest that under certain conditions could compromise the independence of the contribution made by an expert. By definition, all those involved in providing high quality advice must have an interest in the relevant subject, in other words they must have been previously involved in researching or studying the subject. In today's scientific world, that implies necessarily direct or indirect contacts, exchanges and even collaboration with a variety of technical, scientific, industrial organisations producing data or conducting research. Quite often the independence of an expert participating in an assessment is challenged on grounds of his or her previous pronouncements on the same issue in a different context. The EU Court of Justice has clarified in a case of 2010 concerning the recognition of a medicinal product as "orphan medicinal product under Regulation (EC) No 141/2000" (European Court of Justice 2008) that:

> The obligation of impartiality cannot be construed as meaning that there is a legal impediment to an expert being consulted in a procedure for the designation of a medicinal product as an orphan medicinal product solely because he has already given an opinion on the same product in another national procedure carried out in a Member State of the European Union.

This conclusion may be considered relevant for, and extrapolated to the independence of experts involved in the work of the EU bodies on risk assessment matters. Nevertheless, the active involvement of a scientist in advocacy activities on a risk-policy issue may instead cast serious doubts on the appropriateness of allowing his/her participation in what must be, and should be seen as an independent assessment-process.

Finally, scientists are human beings and they bring in their work their ethical, religious, philosophical beliefs that may introduce a bias in their judgment. In practice, it is illusory to expect an absolute guarantee of full independence for each participant in a risk assessment committee, panel or group. Nevertheless, the very fact that the risk assessment is a collective activity, where no single participant may determine the conclusions, provides a reasonable assurance of objectivity and balance for the scientific advice delivered by the EU risk assessment bodies. Pluralism and transparency are in fact the main conditions for ensuring the overall independence of the scientific advice provided by a risk assessment body.

In 2013 the Commission has proposed non-binding *Guidelines on the prevention and management of conflicts of interest in EU decentralised agencies* (European Commission 2013), following the invitation in the 2012 Joint Statement of the European Parliament, the Council of the EU and the European Commission on decentralised agencies. Both the Joint Statement and the Guidelines have a very wide scope, covering agencies of different nature. The guidelines are particularly relevant for agencies providing scientific advice. They are aimed at generalising across the wide range of EU agencies best practices on the management of conflicts of interests, like those described in this book, and already developed, applied and constantly improved by bodies like EFSA and the Commission Scientific Committees. The analysis of the problems related to conflicts of interests as well as the broad lines of the management principles are very similar to those already discussed in this Chapter. The Guidelines leave the establishment of the specific criteria and procedural details to the Agencies themselves, in consideration of their independence and the very diverse mandates and tasks of the numerous bodies to which the guidelines are addressed. Nevertheless, the guidelines insist on the need for communication and exchange of practices between agencies as a way to exchange experiences, improve practices and achieve greater consistency.

8.6.3 Transparency

Transparency is vital for ensuring the full value of risk assessment advice in the management and governance processes. The credibility and acceptance of the scientific advice by the stakeholders depend very much on transparency about the procedures, processes and methods applied. Transparency is also necessary in order to ensure the proper understanding of the meaning of and limitations inherent in the risk assessment advice.

Transparency on Procedures and Working Methods Stakeholders nowadays want to get full information on who is delivering scientific advice to the Institutions. Easy access should be ensured to information on the composition of risk assessment bodies, their working groups, the procedures and criteria for their formation, the expertise and independence of members. Transparency should also be ensured on all the procedural steps leading to the adoption of a scientific opinion, from the establishment of a mandate by the requesting institution to the final publication of the risk assessment report. Transparency on procedures and working methods is very developed across all EU risk assessment bodies. All provide documentation in their web-sites, including in most cases details on their organisation, working programmes and procedures, both general and for each specific activity and scientific opinion.

Transparency in Risk Assessment. A Guidance Document on Procedural Aspects Adopted by the EFSA Scientific Committee A guidance document on transparency in risk assessment has been adopted by the EFSA Scientific Committee, on request from the EFSA management. The terms of reference were aimed at obtaining guidance on:

- process-related considerations, e.g. appropriate stakeholder involvement prior and during the risk assessment, handling, justification or explanation of minority opinion;
- consistent and harmonised documentation;
- a sufficiently detailed description of the strengths, robustness and limitations of the data used for the risk assessment;
- a clear description of the underlying assumptions and uncertainties providing the reasoning for decisions;
- a list of criteria for inclusion or exclusion of available scientific information for a given risk assessment, e.g. criteria for selection of pivotal studies and data, being part of the risk assessment;
- structured and stepwise approaches in hazard and risk assessment, e.g. science-based decisions for the need of additional studies based on previous studies in a stepwise approach, resulting in an optimal set of toxicity tests.

In a first report, the EFSA Scientific Committee addressed the procedural aspects. The guidance covers several aspects (EFSA 2006):

- Selecting qualified scientists to participate in EFSA's activities and making sure of their independence
- Overall handling by EFSA of requests for a scientific opinion or for technical and scientific assistance
- Scientific opinions and other types of EFSA documents
- Ensuring availability of relevant data including those from literature
- Information exchange between Scientific Committee or Panel and the originator of the request (i.e. European Commission, European Parliament and the Member States)
- Involvement of stakeholders other than the European Commission, the European Parliament and the Member States
- Dealing with diverging scientific opinions among different bodies
- Adoption of scientific opinions
- Dissemination of scientific opinions, other documents and access to underlying data
- Confidentiality
- Revision and update of scientific opinions

The issues addressed and the considerations made are very similar to those developed in our discussion on best practices. The document identifies several areas for improvement and further work. Interestingly, the guidance of the EFSA Scientific Committee takes the form of internal recommendation adopted without involvement of the EU risk managers.

Transparency on the Scientific Processes: Risk Assessment Policy We have extensively discussed in previous chapters the influence of models, assumptions, extrapolations and uncertainty treatment on the results of a risk assessment. Since risk assessment is not pure science and includes choices and approximations involving a certain degree of subjectivity, a "risk assessment policy" setting the approach on those aspects that could not be considered "science" (for instance scope of the risk assessment and aspects to be addressed, data acceptability criteria, criteria for the defaults to be used, weight of evidence criteria, margins of safety, criteria for "acceptable risk" etc.) should be defined in collaboration between managers and assessors.

The concept of Risk Assessment Policy has been discussed in many documents and articles. It was mentioned at an early stage in the US-NRC "Red Book" of 1982 that noted that in each step of risk assessment (US NRC 1982) there were decisions to be made where risk to human health could only be inferred from the available evidence and the selection from among possible inferential bridges involves both scientific judgements and policy choices. The Red Book introduced the term risk assessment policy to "differentiate those judgements and choices from the broader social and economic policy issues that are inherent in risk management decisions".

An authoritative reference for the concept is also found in the "Working Principles for Risk Analysis for Food Safety for the Application by Governments", formally adopted in July 2007 by the Codex Alimentarius Commission of the FAO-WTO. The working principles prescribe that risk managers provide their risk assessors with explicit risk assessment policies prior to the start of their deliberations. The main elements of a Risk Assessment Policy are (Codex Alimentarius Commission 2007):

- Determination of risk assessment policy, which should be included as a specific component of risk management,
- Risk assessment policy should be established by risk managers in advance of risk assessment, in consultation of risk assessors and all other interested parties. This procedure aims at ensuring that the risk assessment is systematic, complete, unbiased and transparent.
- The mandate given by risk managers to risk assessors should be as clear as possible,
- Where necessary, risk managers should ask risk assessors to evaluate the potential changes in risks resulting from different risk management options.

A risk assessment policy should cover substantive, procedural and interpretative aspects. So far, no jurisdiction has established a formal and comprehensive approach to the definition of risk assessment policies, along the lines proposed by the Codex Commission, across the different sectors of health and environmental protection. Within the Codex system itself, the establishment by risk managers of risk assessment policies has met with resistance and made little progress. Nevertheless, on the EU level, some substantive and procedural aspects have been defined by regulation on a sector-specific level, for instance by prescribing the scope and aspects to be covered by the assessment or the generic principles and criteria to be applied, notably on independence and transparency. Much more has been addressed at technical level, by the risk assessment bodies themselves, in the form of internal procedures or guidelines, but without the formal involvement and endorsement of policy makers. Some more subtle, but very relevant methodological aspects have however escaped any form of harmonisation and endorsement by risk managers and are left to the practice of individual committees or panels, and to the internal initiatives of the risk assessment bodies.

In the second part of its guidance document, published in 2009, the EFSA Scientific Committee has addressed aspects of transparency related to the scientific process that overlap with the main themes of a generic risk assessment policy. That guidance document represents a good example of an internal policy, independently established by the risk assessors in the absence of a comprehensive and formal risk assessment policy, aimed at ensuring transparency on all aspects of the risk assessment process. A further step has been made by the European Commission, incorporating guidelines for the risk assessment process into the rules of procedure of its Scientific Committees. Although the rules of procedure of the Committees are adopted by the Committees themselves, they are proposed by the Commission and represent a good example of approach agreed between managers and assessors.

The EFSA Guidance document provides an important contribution towards a best-practice risk assessment policy. The main issues identified in that document are (EFSA 2009):

General

The scientific outputs must be transparent with regard to the data, methods of analysis and assumptions that are used in the risk assessment process;

- Transparency is needed in all parts of the risk assessment
- To be transparent, a risk assessment should be understandable and reproducible;
- Where possible, harmonised assessment terminology should be used, preferably based on internationally accepted terminology;
- The procedure by which a risk assessment is completed needs to be based on accepted standards of best practice;
- When circumstances require that a scientific assessment is provided within a limited time period (e.g. in a crisis situation), the effect of this on the uncertainty of the response should be explained, and options and timescales for reducing that uncertainty should be described.

Scope and objectives

- The scope and objectives of the risk assessment should be considered and documented, and if necessary, clarified with the originator of the request before the commencement of the risk assessment.

(continued)

Data and data sources

- A risk assessment requires a comprehensive description of the data considered and the experimental and/or environmental conditions under which those data were generated;
- The sources of all data used for the assessment, including unpublished data and personal communications, must be referenced and the data evaluated to determine their quality and relevance to the assessment. These should be reflected in the relative weight given to them in the assessment and taken into account in the overall evaluation of uncertainty;
- If data bases, data banks and information systems are used, their identity should be documented along with the key search terms and strategies applied and time period covered.

Inclusion and exclusion of data

- All the data and information available for the assessment are evaluated but only those judged to be relevant should be used as the basis for risk assessment. The inclusion/exclusion criteria applied to the data should be explained and described within the risk assessment. If data are excluded, this should be stated along with the rationale for their exclusion.

Confidential data

- The requirements in the 2005 Decision of the EFSA's Management Board apply to transparency and confidentiality of data. The approach taken is that the maximum amount of information linked to EFSA's activities shall be disclosed or made accessible to the public, and that only the essential minimum should be kept confidential where this is justified; in particular, the information related to the industrial property which is included in the authorisation dossiers.

Assumptions

- All assumptions should be documented and explained. Where alternative assumptions could reasonably be made, the related uncertainties can be evaluated together with other uncertainties (see below).

Assessment

- In qualitative assessments, conclusions are expressed in a narrative. In quantitative assessments conclusions are based, at least partly, on calculations or mathematical models. In both cases, transparency requires that every element of the reasoning, calculation or mathematical modelling should be communicated and justified;
- Whenever mathematical models are used, it should be described whether, by what means and to what extent they have been validated or evaluated.

(continued)

Variability and uncertainties

- There may be differences in risk due to variability among individuals, populations, species or ecosystems. It is important to identify and describe the most influential contributors to variability in risk, preferably by statistical analysis of the underlying data. Any statistical difference must be interpreted in the light of its biological relevance;
- Although it may be impossible to identify all the uncertainties, each scientific output should describe the types of uncertainties encountered and considered during the different risk assessment steps, and indicate their relative importance and influence on the assessment outcome;
- Expression of uncertainty and variability in risk estimates may be qualitative or quantitative, but should be quantified to the extent that is scientifically achievable;
- Where factors are used to account for uncertainty, an explanation of their basis and their appropriateness or a reference to documents where that information may be found should be included;
- Where point estimates are used for variable or uncertain quantities, justification for the values chosen and assessment of their influence on the assessment should be included.

Conclusions of a scientific opinion

- Conclusions should address the terms of reference, should reflect the scope and objectives of the risk assessment and provide characterisation of the risk under consideration. All key scientific information underpinning the assessment should be outlined including uncertainties and data gaps;
- Conclusions should be based only on data previously described;
- The reasoning leading to the conclusions should be described. Where applicable, the degree of consistency with risk assessments by other bodies should also be stated.
- Risk assessments may be performed on a particular compound, agent or topic by different risk assessment bodies at national, European or international level. Such opinions should be considered by EFSA. Their relevance to EFSA's own risk assessment should be evaluated provided that a comprehensive description of all data, processes and methods is available;
- The same data set may, however, not be appropriate in a different context. Therefore, the terms of references need to be checked carefully before considering whether an opinion expressed by another body/committee can be used by EFSA;
- In case of diverging opinions, the procedure foreseen in Article 30 of Regulation (EC) No 178/2002 should be followed closely to identify and possibly to resolve diverging scientific opinions.

In conclusion, the minimum requirement on transparency is that scientists document the processes, methods, assumptions, defaults and criteria used and explain in the risk assessment report their influence on the outcome. Uncertainty and variability should in particular be well documented and discussed in a manner easily comprehensible by managers. In the end, transparency on processes, methods, etc. should ensure full reproducibility of a risk assessment. It should also allow understanding the reason for possible divergent opinions on the same subject (different data sets used, different models, assumptions or methods etc.). However, the aim to document in the risk assessment reports details of the process and methods applied must be balanced with the need for practicability and for preserving readability by risk managers and stakeholders.

There may also be some tension between transparency and the objective of preserving independence of the risk assessment process. For instance, early disclosure of the names of working groups' members as well as the agendas of their meetings may expose the participant scientists to pressures from lobbies offering to meet and brief them, to provide documents, data, position papers etc., in an attempt to influence them through somewhat opaque, behind the scene contacts. When risk assessment bodies deal with very sensitive and controversial subjects, a certain degree of confidentiality and specific protection measures may be necessary and justified.

8.7 Conclusions and Recommendations of the Commission Scientific Committees for the Improvement of the Utility of Risk Assessment

In their already opinion mentioned on "Making Risk Assessment More Relevant for Risk Management", the three Scientific Committees of the Commission have offered some useful conclusions and recommendations that fit well with the analysis provided in this Chapter:

- Key messages from the risk managers and policymakers were that the outputs of risk assessment needed to be more policy and management relevant and this ought to be facilitated by more dialogue.
- Given that management decisions are often taken against a backdrop of trade-offs between the benefits of interventions for human health and environment and the costs of restrictions on other parts of societal welfare, it follows that to be "management relevant" risk assessments need to inform these cost/benefit analyses.
- A similar problem is encountered in integrated risk assessments where it is often necessary to compare impacts across very different entities, such as humans and ecosystems. Such comparisons can only be achieved by

(continued)

weighting on the basis of people's preference if social-decision rules are to be based on democratic decision-making – and one expression of public preference is economic value – so the risk assessment outputs need to be compatible with such valuation.

- In pursuing Term of Reference 3, "identifying risk assessments that …are informative", the Scientific Committees therefore focused on the extent to which risk assessments are compatible with socio-economic analyses,… The Scientific Committees concluded that there is considerable confusion on the needs that socio-economic analysis put on risk assessment and that to be more useful …risk assessments should be expressed in terms of value-relevant impacts on humans and ecosystems rather than on the somewhat technical surrogates often included in the routine risk characterizations …
- A key recommendation was that the European Commission Scientific Committees be invited to express Opinions on risk in terms of likely impacts on human health and ecosystems services expressed to the extent possible in probabilistic and value-relevant terms.
- To make this change, the Scientific Committees recommend more dialogue between risk assessors and socio-economists.
- Expressing risk in terms that matter for regulators will also facilitate communication – but to enhance that the Scientific Committees make recommendations on a system of dialogue that facilitates the exchange of information between risk assessors and risk managers while ensuring the scientific integrity of the of the risk assessment
- The Scientific Committees also recommend extending this dialogue to all stakeholders in both initial forums and final consultations as a way of clarifying issues and ensuring more buy in. This will be especially important where the issues are complex and the outcomes are likely to be of major socio-economic importance.
- Recommendations were also made for improving risk assessment reports, in particular in terms of: the evaluation of different possible scenarios; full characterization of the whole populations/ecosystems at risk with attention to particularly sensitive subpopulations/species; clear expression of uncertainty; explicit disclosure of hypotheses used without supporting evidence.
- One step in facilitating the development of better practice could be to arrange training for both assessors and managers based on a common manual – and Making Risk Assessment More Relevant for Risk Management …could provide a starting point for that.
- Throughout there are important concerns about uncertainty. It impinges on all aspects of risk assessment and socio-economic analyses. It needs to be made transparent…

Summary

1. The starting point for improving the utility of risk assessment is to ensure a better understanding by all the actors and stakeholders, across all the risk analysis stages, of both the potential and the limits of science in determining the reality, probability, severity, extent and distribution of risks.
2. The full added value of science-based risk assessment cannot materialise in the decision making process if risk managers and decision makers are not made aware of the nature and limits of the assessment tools and are not in a position to properly interpret and use the results of an assessment. To that aim, risk assessment reports should transparently present all important assumptions underlying the conclusions reached as well as the limitations of such conclusions.
3. It happens that health and environmental risks are invoked to justify measures pursuing different public policy objectives or intended to respond to democratically expressed public concerns or preferences. Policy makers sometimes tend to "hide" behind science to take decisions that are motivated by considerations that have little to do with scientific evidence on health, safety and environmental protection. There is a need for a more comprehensive, transparent and internationally recognised framework of criteria for the assessment of policy options.
4. Scientists should be careful to show that they accept that science alone cannot decide the "right" course of action, irrespective of economic and political aspects, interests and values of the parties involved and the other relevant societal aspects. They should refrain from entering the policy debate, where values, costs, benefits, interests play a legitimate and important part.
5. Risk managers try to maximize the legal security of the measures that they take by adhering in their mandates to risk assessors to the exact wording of the relevant regulatory provisions. However, this type of direct link between assessment and management is inappropriate on two grounds: often it pushes scientists to oversimplify the expression of risk (for instance to reply with a "safe, not-safe" or similar statement); moreover, it quite directly involves scientists in risk management.
6. Risks should be evaluated together with costs and benefits for the various possible regulatory options. Again, it is necessary to that aim to avoid a too rigid separation between risk assessment and management: a systematic dialogue and a close collaboration between the assessment bodies and managers should take place across the various stages of policy analysis, while preserving the integrity of risk assessment.
7. Sometimes, risks are expressed in ways that policy makers may find it difficult to understand or interpret. Policies and specific measures addressing risks are intended to protect goods that society values, in particular health, the environment, biodiversity or property. Nevertheless, when risks are expressed in technical terms, it is difficult for decision makers to obtain from such expressions a clear picture of what is at stake and to see what benefits, in terms of protection of social goods, could be achieved by the options considered.

8. Transparency about uncertainty and variability is an essential requirement of best-practice risk assessment. Various approaches and methods have been defined and are applied by the EU risk assessment bodies to manage uncertainty. Nevertheless no consistent approach is available for now across sectors for the expression of uncertainties. A sustained dialogue should take place between assessors and managers on how to present uncertainties, which expressions and metrics would best meet the needs of managers and how to minimise the risk of misunderstandings on the interpretation of uncertainties.
9. The EU has set excellence, independence and transparency as core principles of its reorganized risk assessment system. Additionally, the Commission "Better Regulation" initiative has enhanced the focus on responsiveness and openness in all policy and regulatory areas. These principles have been incorporated into the rules and practical functioning of the EU risk assessment bodies.
10. Independence includes both the independence of science from policy considerations and political interests, as well as the separation between risk assessment and risk management responsibilities, and the independence from any vested interest of lobbies and the distance from political and advocacy positions.
11. Transparency involves both the transparency of the procedures and working methods applied and the full, pro-active disclosure and communication of the scientific processes, data, assumptions, methods and criteria used for reaching the risk assessment conclusions. Transparency also requires disclosure of minority opinions.
12. The objectives of openness and responsiveness have been put in practice through procedures for pro-active public consultation and stakeholder dialogue, with a view to get all useful information helping to shape the questions to risk assessors and to address the scientific issues in the most appropriate way in order to respond to stakeholders' questions and concerns.
13. Ensuring scientific "excellence" is a complex exercise that does not stop with the selection of members of a scientific committee. The issues on which high quality scientific advice is most needed are often complex, multidisciplinary, new or emerging, and are frequently characterised by uncertainty and scarcity of pertinent data for risk assessment. "Competence" can only be ensured by providing for adequate collective expertise, including through the ad hoc involvement of experts external to the relevant standing committees or panels.
14. In ensuring the objective of independence, a distinction must be done between the presence of an interest and a conflict of interest that under certain conditions could compromise the independence of the contribution made by an expert. By definition, all those involved in providing high quality advice must have an interest in the relevant subject, in other words they must have been previously involved in researching or studying the subject. In today's scientific world, that implies necessarily direct or indirect contacts, exchanges and even collaboration with a variety of technical, scientific, industrial organisations producing data or conducting research. It is illusory to expect an absolute guarantee of full independence for each participant in a risk assessment committee, panel or group. Nevertheless, the very fact that the risk assessment is a collective activity,

where no single participant may determine the conclusions, provides a reasonable assurance of objectivity and balance for the scientific advice delivered by the EU risk assessment bodies. Pluralism and transparency are in fact the main conditions for ensuring the overall independence of the scientific advice provided by a risk assessment body.

15. Stakeholders nowadays want to get full information on who is delivering scientific advice to the Institutions. Easy access should be ensured to information on the composition of risk assessment bodies, their working groups, the procedures and criteria for their formation, the expertise and independence of members. Transparency should also be ensured on all the procedural steps leading to the adoption of a scientific opinion, from the establishment of a mandate by the requesting institution to the final publication of the risk assessment report.
16. Since risk assessment is not pure science and includes choices and approximations involving a certain degree of subjectivity, a "risk assessment policy" setting the approach on those aspects that could not be considered "science" (for instance scope of the risk assessment and aspects to be addressed, data acceptability criteria, criteria for the defaults to be used, weight of evidence criteria, margins of safety, criteria for defining "acceptable risk" etc.) should be defined in collaboration between managers and assessors. A risk assessment policy should cover substantive, procedural and interpretative aspects. So far, no jurisdiction has established a formal and comprehensive approach to the definition of risk assessment policies.

References

Codex Alimentarius Commission. (2007). *Working principles for risk analysis for food safety for the application by governments.*

EFSA (European Food Safety Authority). (2006). Transparency in risk assessment carried out by EFSA: Guidance document on procedural aspects. *The EFSA Journal, 353*, 1–16.

EFSA (European Food Safety Authority). (2009). Transparency in risk assessment-scientific aspects. *The EFSA Journal, 1051*, 1–22.

European Commission. (1997). Consumer health and food safety. *COM* (97) 183.

European Commission. (2002). The collection and use of expertise by the commission. *COM* (2002) 713.

European Commission. (2013). *Guidelines on the prevention and management of conflict of interest in EU Decentralized Agencies.*

European Court of Justice. (2008). *Case T-74/08*

Gluckman, P. (2014). The art of science advice to government. *Nature, 507*, 163–165.

IPCC (International Panel of Climate Change). (2010). *Guidance note for lead authors of the IPCC fifth assessment report on consistent treatment of uncertainties.*

SCCS, SCHER, SCENIHR. (2013). *Making risk assessment more relevant for risk management.*

US NRC (National Research Council). (1982). *Risk assessment in the federal government: Managing the process.* Washington, DC: National Academies Press.

US NRC (National Research Council). (2009). *Science and decisions.* Washington, DC: National Academies Press.

Chapter 9
Summary of Key Issues and Best Practices in Risk Analysis

9.1 Introduction

This Chapter follows up on the previous ones and provides a synthetic list of key aspects for a best-practice approach to risk analysis, together with some guidance on each aspect. Although in previous Chapters we have also briefly discussed certain aspects and problems of the EU regulatory policy and practice in order to provide a broader perspective on the approach to risk, we focus here on the main theme of the book, risk analysis. The summary of best practices presented here does not aim to cover all the details of the organisation and operation of the different components of risk analysis. The summary proposed should be seen in light of the more detailed discussion of the corresponding issues presented in this book. The issues addressed have been identified examining the guidelines, the rules and practices of the EU Institutions and risk assessment bodies, as well as the practices in other jurisdictions, and are the basis for establishing a high quality risk analysis approach. Obviously, the application of these practices needs to be modulated and adapted depending on the nature and complexity of the risk issues considered, the practical structures in place for risk assessment and risk management and the applicable legal framework. This summary of best practices may help establishing check lists for supporting the quality assessment of current or proposed procedures and practices.

9.2 An Extended and Improved Risk Analysis Paradigm

- **Full, accessible information to stakeholders and the public on the organisation, procedures and activities of the scientific risk assessment process**

B. Delogu, *Risk Analysis and Governance in EU Policy Making and Regulation*,
DOI 10.1007/978-3-319-30822-7_9

Procedures should be put in place and pro-actively communicated in order to ensure full information to stakeholders and the public, and transparency on the way in which risk assessment is organised and conducted. This includes in particular, information on the composition of the scientific bodies, committees and groups, how the members are selected, how transparency on any interest is ensured and conflicts of interests are dealt with, the work programme, minutes of meetings, planned consultations and access to the resulting documents.

- **Pre-assessment and framing of risks, including concern assessment**

Before launching a risk assessment, a dialogue should take place between risk managers, assessors, stakeholders and experts in order to explore the interests, concerns, perceptions, societal aspects, values and expectations at stake. Experts should provide a preliminary description of the state of knowledge and the identification of problem areas and data needs.

- **Preliminary definition of policy options**

In view of a comprehensive mandate to risk assessors, the main policy options should be tentatively identified by risk managers at an early stage. Assessors would then be in a better position to optimise the utility for risk management of their scientific assessment and to advise on the risk reduction potential of the options.

- **Multi-actor and multi-stakeholder dialogue and collaboration aimed at the pre-assessment of risks and the definition of questions for scientific risk assessors**

The stakeholder dialogue for the risk pre-assessment should be further developed with a view to ensuring that the formulation of the risk-issue and the questions submitted to risk assessors properly address all the relevant problems and concerns, reinforcing the relevance of scientific advice for the decision making and governance process.

(continued)

Depending on the nature of the issue and the applicable legal framework and decision procedure, policy makers should be informed of the envisaged consultation of risk assessors, the questions to be put and the results of the concern assessment, in time for them to provide an input as appropriate.

- **Risk managers-risk assessors dialogue in view of launching the risk assessment**

Risk managers should present the issue and the questions to risk assessors in a dialogue aimed at ensuring that risk assessors get a full understanding of the objectives of risk managers and a clear view of the relevant background, risk-framing and concerns. This dialogue may need to be reiterated once risk assessors have preliminarily explored the issue.

- **Scientific risk assessment: independent from political constraints and vested interests, but not conducted by scientists in isolation**

Scientists conducting the risk assessment should be independent of any influence, political, economic, ideological etc. Nevertheless, they should not work in complete isolation. The dialogue with risk managers may need to be repeated during the course of the assessment in order to optimise the utility for risk managers of the final result. Risk assessors should also have opportunities to consult external experts, scientific bodies and researchers.

- **Open call for data and information, technical hearings and/or expert workshops, as appropriate, with appropriate follow up**

Depending on the nature of the issue and the state of knowledge, it may be necessary to collect data and information through open calls. The opportunity of such calls should be evaluated in light of both the needs of assessors and the expectation of stakeholders to see all information that they may consider relevant to be examined by the scientists. Technical hearings and/or expert workshops may be appropriate in order to bring-in useful information and expertise and to allow expression of diverse views and evaluations of the relevant issues. The way in which data, information and expressed views have been considered should be explained in the final risk assessment report.

- **Joint examination by risk managers and scientists of risk assessment policy issues arising in the technical assessment**

Scientists should consult with risk managers on issues involving value judgement, subjective assumptions or criteria and all the other aspects which usually go under the category of "risk assessment policy issues". This consultation should be framed within a general risk assessment policy endorsed by the policy makers. The risk assessment report should transparently explain how such issues have been dealt with.

- **Public consultation on the preliminary opinion resulting from the scientific risk assessment (it may include stakeholder hearings)**

A further public consultation, which may include a stakeholder hearing, should take place once a preliminary risk assessment opinion is published. The objective is to collect comments, questions, suggestions and criticisms useful for establishing a final improved version properly responding to any remaining concerns.

- **Final dialogue risk assessors-risk managers on the results of risk assessment**

A final dialogue should take place between assessors and managers on the results of risk assessment, aimed at their full understanding by managers, allowing for any necessary clarification and complement where necessary, with particular focus on uncertainties and their consequences on the estimated risk and its distribution, the limitations due to the methods used and the availability and quality of data, the expression and grading of risk, the possible risk-risk balance and the reasons for diverging opinions or minority positions.

9.3 A High Quality Risk Assessment Approach

- **Risk Assessment Policy**

Risk managers should define a risk assessment policy in collaboration with risk assessment bodies, in light of the already established experience, guidelines and best practices. The policy should cover all aspects of the risk assessment process and activities others than the purely scientific concepts and methods, notably the issues related to the excellence, independence and transparency of the risk assessment advice. The policy should be endorsed by the EU political institutions responsible for the decisions on health, safety and environmental protection.

- **Open-call selection of scientists and experts**

Scientists and experts performing risk assessments should be selected on the basis of open calls for expression of interest, pre-defined criteria on excellence and independence, a transparent and neutral selection process not influenced by any political, economic or national interests.

- **Multi-disciplinary and pluralist composition of scientific risk assessment bodies**

Risk assessment bodies should involve a wide range of disciplines, different profiles from academy and research and experts from different professional horizons, with a view of ensuring a pluralistic disciplinary and cultural approach. The objective is to prevent the dominance of a single school of thought on problems characterised by rapid evolution of scientific knowledge and interdisciplinary complexity. Gender and geographical balance should also guide the selection of scientists and experts, as far as possible and compatible with the overarching principles of excellence and independence. Transparent procedures should be in place to allow access to expertise provided by industry experts while preserving the independence and integrity of the scientific advice.

- **Measures aimed at involving the most qualified scientists, researchers and experts**

In order to ensure the participation of the most qualified scientists, experts and researchers, adequate resources should be allocated covering the cost of participation and allowing an adequate compensation of the participants and/or their institutions or bodies; working methods and appropriate infrastructure should be established allowing for participation from distance in order to reduce waste of time. The contribution of scientists and experts should be visible, recognised and given an adequate and visible public value.

- **Procedures and resources to provide high quality technical support to risk assessors**

In addition to adequate administrative and logistic support, scientist should have access to high quality technical support for data collection and elaboration, literature search, drafting, editing etc. Such procedures should ensure that the resulting scientific advice is delivered in light of all the relevant scientific and factual data as well as a full review of the state of knowledge.

- **A clear, shared understanding of the respective role of scientists, risk managers and decision makers**

All the actors in the risk analysis chain should be aware and respectful of the respective roles. Risk managers and decision makers should not interfere with the independence of scientists and should not attribute to "science" the responsibility of their decisions. Policy makers and risk managers should assume full responsibilities, within the framework of their respective roles and powers, for the quality of the entire process leading to final decisions on the political level. Risk managers should not directly or indirectly ask risk management questions to scientists. Questions should address risks, not overall pronouncements on safety of products, technologies etc. Scientists in turn should refrain from taking position on risk management, should not comment publicly on decisions concerning issues in which they have been involved as assessors. Scientists should be aware and accept that decisions are taken by political institutions, accountable to the public, that act in light of a multi-criteria appraisal in which "science" is just one of the relevant elements. Scientists actively advocating a particular risk management or policy position should not be involved in risk assessment on the issues in question.

- **Measures to ensure advice free from undue influences**

Appropriate procedures should be established in order to protect the integrity of scientific risk assessment by ensuring transparency on interests of all those intervening in the process and applying measures to prevent or manage conflicts of interest, requiring full collegiality in the decisions and full transparency on methods, assumptions, weight of evidence criteria, data collection and selection. Risk assessment activities should be organised in a way preventing a single line of thought to gain control of the process. Credible procedures and means should be put in place and communicated aimed at identifying and managing conflicts of interests.

- **Transparent, stable and consistent assessment methods, assumptions, defaults, criteria**

Methods, assumptions, defaults and criteria should not be decided on an ad hoc basis. Subjectivity should be limited by establishing horizontal and stable standards to be consistently applied, and requiring full justification for deviations. The standards applied and their rationale should be clearly explained.

- **Dialogue and collaboration with all relevant scientific bodies, and procedures to resolve or explain divergent opinions**.

Dialogue and collaboration between the EU risk assessment bodies and, as necessary, with national and non-EU bodies, should aim at improving the scientific quality of advice by sharing best-practices. promoting methodological consistency and working together on issues of common interest. Although each risk assessment body should act in an independent manner, in case of divergent opinions on the same subject, pro-active efforts should be made in order to resolve or at least explain the divergences. Risk managers should request risk assessment bodies to motivate in a way understandable to managers and policy makers the source and reasons of divergent opinions (different data sets, different interpretation of data etc.). It is the responsibility of risk managers to resolve divergences and report to policy makers on them.

- **Procedures for a sustained dialogue with risk managers and procedures for stakeholder dialogue and stakeholder/public consultation, while preserving the independence and integrity of risk assessment**

The extended risk analysis paradigm includes a closer and sustained dialogue between managers and assessors as well as dialogue and consultation with stakeholders and the public. Risk assessment bodies should establish and communicate procedures for allowing such dialogues in an ordered, structured and transparent manner, while preserving the independence of risk assessment.

- **Expression of risk related to goods that society values, in terms useful for risk management needs**

Risks should be expressed in terms and with metrics that relate to the protection objectives of risk managers and to what the public values, for example loss of quality-adjusted life-years, additional incidence of cancer, loss of biodiversity. The choice of the appropriate expression and metrics should be discussed between assessors and managers.

- **Full, clear and understandable reporting of methods, assumptions, criteria, uncertainties**

Risk assessment reports should include clear and understandable explanation of the methods, assumptions and criteria used as well as of the uncertainties and variability and their impact on the conclusions. Sensitivity analysis should be used, as appropriate, in order to allow understanding the extent of variability of the risk in function of the assumptions, uncertainties and variability. Guidance should be developed on how uncertainties should be expressed in order to support the appraisal of risk by managers and policy makers, ensuring consistency across sectors and risk assessment opinions. The comprehensive description of the reasoning, methods, data and assumptions should in particular allow the reproducibility of the assessment and a comparison with assessments leading to divergent results. Standards for reporting on such aspects should be part of the risk assessment policy. In general, risk assessment reports should be written with the needs of managers and policy makers in mind.

- **Clarity on the limitations of a risk assessment**

Risk assessors should highlight to risk managers the possible limitations of their risk assessment in cases where cumulative exposure from different sources to the same substance or stressor is not taken into account and may be significant in determining the actual resulting risk. Cumulative effects from combined exposure to different substances and/or stressors should be taken into account as far as possible and risk assessors should state clearly whether and how such effects have been taken into account, what alternative methods could have been applied, and the possible relevance of and limitations from the way in which cumulative effects are/are not considered in the risk assessment.

9.4 A Comprehensive Approach to Risk Management

- **A full consideration of the different dimensions of risk and of the multi-faceted societal interests and values**

A participatory approach should be applied leading to the factoring of the complete set of interests and societal values in the identification and assessment of options. A real multi-criteria assessment should be performed, avoiding the artificial reduction of all negative effects and potential benefits to the sole monetary dimension. Clarity should be made on the imperative requirements and other legitimate objectives and factors pursued, and a distinction should be maintained in the justification of measures between risk-related and other considerations.

- **A systematic factoring in of uncertainty, variability and distributional aspects**

The assessment of the different risk management options should take into account the impact of uncertainties, variability and the distribution of risks and benefits across the various groups.

- **An objective assessment of the state of knowledge and the uncertainties involved that may require the application of the precautionary principle**

Risk managers should assess the quality, robustness and reliability of the scientific knowledge and data used for a risk assessment, the level and implications of remaining uncertainties, they should consider carefully divergent or minority opinions. In light of such assessment, risk managers should examine whether, to what extent and how the precautionary principle needs to be taken into account in determining the appropriate measures. When scientific consensus is not achieved, there is nevertheless a duty on the Institutions to decide on the line of reasoning to follow and the appropriate course of action.

- **A comprehensive, proportional and adaptive definition of policy options, in consideration, in particular, of the proportionality principle**

In determining the measures to be taken, the proportionality principle should be taken into account in order to ensure the most efficient course of action. Options should include measures for the monitoring and revision of regulation, particularly where the Precautionary Principle is applied. Should be explored possibilities for measures justified by ancillary benefits and measures that can be adapted in light of new knowledge and evidence and of the results from the monitoring of their effectiveness and impacts.

- **A proper use of the Precautionary Principle (PP)**

Use of the Precautionary Principle should follow consistently the guidelines established by the European Commission. The starting point should always be a review of the state of knowledge made by independent scientists. The integrity of the PP should be preserved. The PP should not be invoked to justify measures motivated by requirements or objectives that are not related to risk aspects. Measures taken under the PP should be as far as possible adaptive, reversible, proportionate and subject to monitoring. When the application of the PP is in line with the above-mentioned guidelines, calls for "more science" and other delaying tactics should be firmly resisted.

- **Consideration of risk-risk balance**

The ancillary risks and benefits of the options considered should be identified and evaluated, as far as relevant and reasonable. Where necessary, the results of such assessment should guide the design of more comprehensive measures aimed at preventing or mitigating the ancillary risks and enhance the ancillary benefits.

- **A comprehensive consideration of costs and benefits, positive and negative impacts**

Costs and benefits should be established in a comprehensive manner, including both monetary and non-monetary aspects. Costs and other negative impacts of inaction should also be factored in the option assessment. The distribution of costs and benefits should be assessed and reported. Options should not be compared simply based on their net monetary cost-benefit balance when non-monetary aspects are important in order to characterise the options considered.

- **Transparent risk acceptability criteria**

Risk managers should apply transparent criteria for risk acceptability, endorsed or at least shared with policy makers. Risk assessors should refrain from pronouncing on the acceptability of risks independently from agreed criteria. To that aim, risk policy should provide guidance on the risk grading criteria and the appropriate terminology. Terms like for example serious, negligible, high, low risk should not be used without a transparent definition and reference to agreed criteria.

- **A balanced consultation and consideration of all interested parties**

The consultation process on the risk management options should be inclusive. Appropriate measures should be taken in order to ensure a balanced participation of all stakeholders and groups, with some support to the weakest participants to compensate for the different level of resources and expertise among stakeholders.

- **A timely, proportionate application of assessment procedures**

The factor time is of the utmost importance in dealing with health, safety and environmental risks. Risk assessment, risk management and impact assessment procedures are necessary for ensuring the quality and effectiveness of policy and regulatory decisions. Nevertheless, a reasonable balance should be struck between the scope and comprehensiveness of such procedures and the need to preserve the effectiveness and timeliness of the decision making. Financial, administrative, technical resources are limited and devoting an excessive assessment effort to certain measures may have high opportunity-cost and costs for the taxpayer not justified by the marginal benefits achieved.

- **A transparent and objective representation to policy makers of both the scientific risk assessment and the comprehensive multi-criteria assessment of policy options, informing in a comprehensive manner the final political appraisal of risks and of possible management measures**

Risk assessment and risk management should be seen as preparatory steps aimed at informing decision makers in view of political decisions. The results of the assessment performed should be represented in a clear, transparent and objective manner, allowing the decision-making institutions to appraise from the political angle the risk measures in full knowledge of all the scientific, economic and societal aspects of the risk issue considered.

MIX
Papier aus verantwortungsvollen Quellen
Paper from responsible sources
FSC® C105338

If you have any concerns about our products,
you can contact us on
ProductSafety@springernature.com

In case Publisher is established outside the EU,
the EU authorized representative is:
Springer Nature Customer Service Center GmbH
Europaplatz 3, 69115 Heidelberg, Germany

Printed by Libri Plureos GmbH
in Hamburg, Germany